Prehistoric Cultures and Environments in the Late Quaternary of Africa

edited by

John Bower and
David Lubell

Cambridge Monographs in African Archaeology 26

BAR International Series 405
1988

B.A.R.

5, Centremead, Osney Mead, Oxford OX2 0DQ, England.

GENERAL EDITORS

A.R. Hands, B.Sc., M.A., D.Phil.
D.R. Walker, M.A.

BAR S405, 1988: 'Prehistoric Cultures and Environments in the Late Quaternary of Africa'

© The Individual Authors, 1988

The authors' moral rights under the 1988 UK Copyright,
Designs and Patents Act are hereby expressly asserted.

All rights reserved. No part of this work may be copied, reproduced, stored, sold, distributed, scanned, saved in any form of digital format or transmitted in any form digitally, without the written permission of the Publisher.

ISBN 9780860545200 paperback
ISBN 9781407346854 e-book
DOI https://doi.org/10.30861/9780860545200
A catalogue record for this book is available from the British Library
This book is available at www.barpublishing.com

CAMBRIDGE MONOGRAPHS IN AFRICAN ARCHAEOLOGY

General Editor

John Alexander, M.A., Ph.D., F.S.A.

St. John's College
Cambridge CB2 1TP, England

Manuscripts should be submitted to
Dr. Alexander in the first instance.

Titles already published:

1. 'The Niger Delta: Aspects of its Prehistoric Economy and Culture' by Nwanna Nzewunwa. BAR-S75.

2. 'Prehistoric Investigations in the Region of Jenne, Mali' by Susan Keech McIntosh and Roderick J. McIntosh. BAR-S89.

3. 'Off-Site Archaeology and Human Adaptation in Eastern Africa: An Analysis of Regional Artefact Density in the Amboseli, Southern Kenya' by Robert Foley. BAR-S97.

4. 'Later Pleistocene Cultural Adaptations in Sudanese Nubia' by Yousif Mukhtar el Amin. BAR-S114.

5. 'Settlement Patterns in the Iron Age of Zululand: An Ecological Interpretation' by Martin Hall. BAR-S119.

6. 'The Neolithic Period in the Sudan, c. 6000 - 2500 B.C.' by Abbas S. Mohammed-Ali. BAR-S139.

7. 'History and Ethnoarchaeology in Eastern Nigeria: A Study of Igbo-Igala relations with special reference to the Anambra Valley' by Philip Adigwe Oguagha and Alex Ikechukwu Okpoko. BAR-S195.

8. 'Meroitic Settlement in the Central Sudan: An Analysis of Sites in the Nile Valley and the Western Butana' by Khidir Abdelkarim Ahmed. BAR-S197.

9. 'Economy and Technology in the Late Stone Age of Southern Natal' by Charles Cable. BAR-S201.

10. 'Frontiers: Southern African Archaeology Today' ed. M. Hall, G. Avery, D.M. Avery, M.L. Wilson and A.J.B. Humphreys. BAR-S207.

11. 'Archaeology and History in Southern Nigeria: The ancient linear earthworks of Benin and Ishan' by P.J. Darling. BAR-S215.

12. 'The Latest Stone Age of Southernmost Africa' by Janette Deacon. BAR-S213.

13. 'Fisher-Hunters and Neolithic Pastoralists in East Turkana, Kenya' by John Webster Barthelme. BAR-S254.

14. 'The Archaeology of Central Darfur (Sudan) in the 1st Millennium A.D.' by Ibrahim Musa Mohammed. BAR-S285.

15. 'Stable Carbon Isotopes and Prehistoric Diets in the South-Western Cape Province, South Africa' by Judith Sealy. BAR-S293.

16. 'L'art rupestre préhistorique des massifs centraux sahariens' by Alfred Muzzolini. BAR-S318.

17. 'Spheroids and Battered Stones in the African Early and Middle Stone Age' by Pamela R. Willoughby. BAR-S321.

18. 'The Royal Crowns of Kush' by L. Török. BAR-S338.

19. 'The Later Stone Age of the Drakensberg Range and its foothills' by H. Opperman. BAR-S339.

20. 'Socio-Economic Differentiation in the Neolithic Sudan' by Randi Haaland. BAR-S350.

21. 'Later Stone Age Settlement Patterns in the Sandveld of the South-Western Cape Province, South Africa' by Anthony Manhire. BAR-S351.

22. 'L'art rupestre du Fezzan septentrional (Libye) by Jean-Loïc le Quellec. BAR-S365.

23. 'Archaeology and Environment in the Libyan Sahara' edited by Barbara E. Barich. BAR-S368.

24. 'The Early Farmers of Transkei, Southern Africa' by J.M. Freely. BAR-S378.

25. 'Later Stone Age Hunters and Gatherers of the Southern Transvaal' by Lyn Wadley. BAR-S380.

26. 'Prehistoric Cultures and Envvironments in the Late Quaternary of Africa' edited by John Bower and David Lubell. BAR-S405.

Contents

Figures iv
Tables vii
Plates viii
List of Contributors ix

Chapter

1 Holocene Nile Floods and Their Implications for the
   Origins of Egyptian Agriculture, by F. A. Hassan 1

2 Climatic Change and Man in the Sahara, by N. Petit-Maire .. 19

3 After the Deluge: The Neolithic Landscape in North Africa,
   by M. A. J. Williams 43

4 Neolithic Adaptations on the Central Nile,
   by A. Mohammed-Ali and E. El-Anwar 61

5 Holocene Environments and Occupations in the Southern
   Atbai, Sudan, by A. E. Marks and K. Sadr 69

6 Evolution of Stone Age Food-Producing Cultures in
   East Africa, by J. Bower 91

7 Environment and Culture in the Late Quaternary of Eastern
   Africa: A Critique of Some Correlations, by P. Robertshaw . 115

8 Patterns of Environment Utilization By Late Prehistoric
   Cultures in the Southern Congo Basin, by S. F. Miller 127

9 The Scale and Timing of Technological and Environmental
   Changes Over the Last 20,000 Years in the Southern Cape,
   South Africa, by J. Deacon 145

10 Human Adaptations in South Africa During the
   Last Glacial Maximum, by P. Mitchell 163

11 The Pleistocene/Holocene Transition in the Western
   Cape, South Africa: Observations From Verlorenvlei,
   by J. Parkington 197

12 Putting the Wind Up the Smithfield: Seasons of
   Occupation Inferred for Sub-Recent Bushman
   Surface Sites, by G. Sampson 207

Figures

1.1	Location map of Birket Qarun, the Fayum, Egypt	16
1.2	Chart showing variation in lake level of the historical Moeris lake and its ancestral prehistoric lakes	17
2.1	Th/U and ^{14}C ages for the Shati valley lake	31
2.2	Radiocarbon ages (Gif and U.Q.) for the Holocene of malian Sahara	32
2.3	Holocene lakes and northern interior delta of the Niger in northern Mali	33
2.4	Radiocarbon ages for Holocene human populations along the Atlantic coast of the Sahara	34
2.5	Long-term climatic variations over the past 400,000 years and predictions for the future	35
2.6	Annual rainfall and mobile quinquennial means at Tombouctou since 1927	36
3.1	Location of sites illustrated in 3.2 and 3.3	55
3.2	Times of high late Quaternary flood levels in the lower Blue and White Nile valleys, central Sudan, and corresponding fluctuations in lake levels in and around the Nile basin	56
3.3	Late Quaternary hydrological events at selected localities in North Africa	57
3.4	Present-day rainfall zones in North Africa	58
3.5	Present-day vegetation zones in North Africa	59
3.6	First recorded appearance of certain domesticated plants and animals at selected sites in North Africa	60
4.1	Neolithic sites in the Central Nile	68
5.1	Map of the Eastern Sudan	85
5.2	Geomorphological Map of the Southern Atbai; simplied from NASA Landsat imagery; with outline of survey zone superimposed	86
5.3	Stages in the hydrography of the Gash River; hypothetical reconstruction based partially on information from NASA Landsat imagery	87
5.4	Correlation of the Southern Atbai sequence with that of the Central Sudan in the Khartoum area	88
5.5	Site distribution by phase, within the survey zone of the Southern Atbai	89
5.6	Distribution of sites during the Late Kassala Phase by Group	90
9.1	A comparison of the timing of changes in various environmental and technological parameters at Boomplaas Cave (BPA) and Nelson Bay Cave (NBC)	161
10.1	Distribution of archaeological sites dated to the Upper Pleniglacial 25,000–16,000 bp	192

10.2	Distribution of archaeological sites dated to the Late Glacial 16,000–12,000 bp	192
10.3	Distribution of recent LSA research	193
10.4	Approximate position of the coast with a 140 m depression in mean sea level at *ca.* 20,000 bp	194
10.5	Distribution of archaeological sites dated to the Upper Pleniglacial and the Late Glacial in relation to present mean annual surface temperature (°C)	195
10.6	Distribution of archaeological sites dated to the Upper Pleniglacial and the Late Glacial in relation to present mean annual precipitation	196
11.1	Map of Elands Bay	206
12.1	The weather station is located only 29 km from the eastern rim of the study area, and 87 km from its western rim	224
12.2	Windshelter data were gathered during two seasons	224
12.3	The Mean Annual Windrose reflects the duration of two classes of wind velocity, averaged over 5 years	225
12.4	Mean Monthly Windroses show N-WSW dominant in late winter/early spring, and SE-E dominant in summer	226
12.5	Daily extremes of wind velocity and temperature show windy days (mostly NNW-W) are associated with warmer temperatures. Calmer, cooler days are associated with southeasterlies.	227
12.6	High velocity winds blow mainly from the NNW-WSW vectors	228
12.7	Mean monthly duration of winds from NNW-WNW vectors combined.	228
12.8	Contrasts between day (07-18hrs) and night (19-06hrs) wind directions in midwinter	229
12.9	Mean monthly duration of all North wind blowing at night (19-06hrs) during 1978	229
12.10	Mean duration of all winds blowing in days (07–18hrs) when the maximum temperature failed to rise above 10° C	230
12.11	Monthly distribution of cold days (maximum temperature below 10° C)	230
12.12	Hourly windspeeds and temperatures for May 1, 1978 (*left*) and two days later (*right*)	231
12.13	Hourly windspeeds and temperatures for October 5 (*left*) and September 21, 1978 (*right*)	231
12.14	Hourly windspeeds and temperatures for two typical summer days: January 18 (*right*) and 31, 1978 (*left*)	231
12.15	Most sites are protected from 5 vectors	232
12.16	Vector distributions of 1-vector sites	232
12.17	Vector distribution of 3-vector sites	233

12.18	Vector distribution of 4-vector sites	234
12.19	Vector distribution of 5-vector sites	235

Tables

1.1	Dates, in radiocarbon years before present, for Fayum Neolithic and Late Fayum Neolithic (Predynastic) sites	5
1.2	Radiocarbon dates on charcoal in radiocarbon years bp for pre-Neolithic Holocene sites	8
5.1	Radiocarbon Dates from the Southern Atbai	73
5.2	Southern Atbai Site Characteristics by Phase or Group	76
6.1:	Cultural Entities in the East African PN	93
6.2:	Variation in Lithic and Ceramic Densities Among PN Occurrences, Kenya and Tanzania	103
6.3	Major Features of PN Evolution	105
10.1	Archaeological assemblages from South Africa dated to the Upper Pleniglacial 25,000–16,000 bp	165
10.2	Archaeological assemblages from South Africa dated to the Late Glacial 16,000–12,000 bp	167
10.3	Variation in ecological diversity within site territories	172
10.4	List of site names referred to in Fig. 10.1 and 10.2	176
11.1	Changes in the frequency of some items in the Elands Bay Cave sequence	201
11.2	Frequency of bone and shell tools and ochre at Elands Bay Cave	201

Plates

2.I:	The Lake Shati mollusc thanatocoenosis	37
2.IIa:	Holocene lake deposits at Wadi Haijad	38
2.IIb:	Mud-cracks level in lake deposits	38
2.IIIa:	Section showing Late Pleistocene eolian layers under Holocene lake deposits, Sbeita	39
2.IIIb:	Lakelet deposits in interdune trough, Umm el Assel	39
2.IIIc:	Section with successive swamp, lake and crust layers, Sbeita	40
2.IIId:	Swamp deposits, Telig	40
2.IVa:	Burial, Erg Ine Sakane	41
2.IVb:	Grinding artifacts, Erg Jmeya	41
2.Va:	Foraminifera and ostracods in lake silts	42
2.Vb:	*Melania tuberculata, Bulinus truncatus,* and *Biomphalaria pfeifferi* in lake deposits	42
2.Vc:	*Crocodylus niloticus* dermic plates, Hassi-el-Abiod	42
8.I:	Typical terrain of the southern Congo Basin	135
8.II:	Transversal arrowheads	136
8.III:	Foliate points	137
8.IV:	A Tshitolian blade in *grès polymorphe*	138
8.V:	Tshitolian lanceolates	139
8.VI:	Excavations at Cauma Bridge	140
8.VII:	Mwambumba, a Tchokwe maker of gunflints, demonstrating the indirect punch technique with which he replicates the blades and flake-scar patterns found in many Tshitolian assemblages	141
8.VIII:	*Petits tranchets* from Ndinga Kiitu, illustrating the variations in size and form found at this forest site	142
8.IX:	Core axes from Ndinga Kiitu, showing a range of sizes and forms	143

List of Contributors

John Bower
Department of Sociology and Anthropology
Iowa State University
Ames, IA 50011 USA

Janette Deacon
Department of Archaeology
University of Stellenbosch
Stellenbosch 7600 South Africa

El-Sayed El-Anwar
Department of Archaeology
University of Bergen Norway

Fekri A. Hassan
Department of Anthropology
Washington State University
Pullman, WA 99164 USA

David Lubell
Department of Anthropology
University of Alberta
Edmonton, Canada T6G 2H4

Anthony E. Marks
Department of Anthropology
Southern Methodist University
Dallas, TX 75275 USA

Sheryl F. Miller
Pitzer College
1050 North Mills Avenue
Claremont, CA 91711 USA

Peter Mitchell
Donald Baden-Powell Quaternary Research Centre
60 Banbury Road
Oxford, OX2 6PN, England UK

Abbas Mohammed-Ali
Department of Archaeology, Faculty of Arts
University of Khartoum, Khartoum, Sudan

John Parkington
Department of Archaeology
University of Cape Town
Rondebosch, Cape Province, South Africa

John Parkington
Department of Archaeology
University of Cape Town
Rondebosch, Cape Province, South Africa

Nicole Petit-Maire
Laboratoire de Geologie du Quaternaire
CNRS, Case 907, Luminy,
13288 Marseille Cedex 9 France

Peter Robertshaw
The British Institute in Eastern Africa
P. O. Box 30710, Nairobi, Kenya

Karim Sadr
Department of Anthropology
Southern Methodist University
Dallas, TX 75275 USA

Garth Sampson
Department of Anthropology
Southern Methodist University
Dallas, TX 75275 USA

Martin A. J. Williams
Department of Geography
Monash University
Clayton, Victoria 3168 Australia

Preface

This book began as a conversation in the airport bar at Bangor, Maine, in August, 1980. We had both attended the Sixth Biennial Meeting of the American Quaternary Association, held at the University of Maine in Orono, and were due to meet again the following month in Poland at the first Dymaczewo conference on the origins of food production in northeastern Africa (Krzyzaniak and Kobusiewicz 1984). While waiting for a flight, we reflected on a theme shared by the two conferences, namely, cultural responses to long-term environmental change. It occurred to us that, while the issues related to this theme, such as adaptation and migration, are more or less ubiquitous elements in research on African prehistory, few attempts have been made to address the theme holistically on a pan-African scale. Moreover, it seemed to us that work on late prehistoric cultures and environments in selected regions of the continent had developed to the point where far-ranging examples of human-environment interaction might serve as both stimulus and corrective for future research. Thus, we began to formulate the idea of a symposium, modelled on the erstwhile Burg-Wartenstein meetings, that would focus on questions of culture and environment during later periods of African prehistory.

Although we were unable to organize the "free-standing" symposium we had envisioned in Bangor, we did work out a reasonably close approximation to our original idea of bringing together, in a relaxed setting, a small group of scholars actively pursuing research centered on the theme at hand. The context for our symposium was an all-day, special session of the biennial meetings of the Society of Africanist Archaeologists in America (SAAAM) in April, 1984, at Portland State University, Portland, Oregon. With some exceptions, the papers read at the SAAAM symposium are the ones which, in revised form, comprise this volume. Included among the exceptions are papers read by Steve Brandt, William Farrand, Diane Gifford-Gonzalez and David Lubell but not submitted for publication, and papers not read at the symposium but submitted afterwards by Janette Deacon, Peter Mitchell, Abbas Mohammed-Ali and Martin Williams. All participants agreed that the papers should be published as symposium proceedings and the organizers of the session were charged with responsibility for finding a publisher and editing the revised papers. While two of the participants in the symposium (Brandt and Gifford-Gonzalez) who have not contributed to this volume were prevented from doing so by other commitments, the absence of papers by Farrand and Lubell reflects special circumstances.

Farrand's paper was on the "Environmental setting of Capsian and related occupations in the high plains of northeastern Algeria," while

Preface

Lubell's was titled "Human-environmental relationships during the Holocene in the Maghreb." Both were preliminary statements based on research conducted in Algeria between 1973 and 1980. Changes in Algerian governmental policies have prevented continuation of this research, analyses of the data collected have been delayed (while Lubell waited in the vain hope of a change in policy) and results are incomplete. Consequently, Lubell and his colleagues have been reluctant to publish here either repetition of material already in print (Lubell 1984a, b; Lubell *et al.* 1982–83; Lubell *et al.* 1984) or statements on data as yet incompletely understood. Having now given up all hope of continuing the project, they are preparing a final report using the data available, and this will be published in due course as a monograph.

A word about the scope and intent of the present volume is in order. The cultural geography of Africa, with its extraordinarily varied array of societies adjusted to an equally complex mosaic of environments and its history of repeated social crises born of climatic deterioration, implies that environment is not only a powerful but also an obvious determinant of human history within the continent. Such a perception has undoubtedly contributed to the frequent occurrence of the culture-environment theme in prehistoric research on Africa. While recent studies of interaction between culture and environment in later periods of African prehistory have often revealed broad parallels with modern cultural geography, they have also shown that the relationship may be subtle, even elusive, and that its effects may be observable at scales ranging form the camp site to the region. The papers included in this volume are intended to expose some of the potentialities and limitations of research on prehistoric cultures and their environments through a series of case studies representing work in various parts of Africa. The concern here is with breadth rather than depth; thus, the volume should be viewed as a sampling of culture-environment research in Africa, rather than an exhaustive treatment of the subject. Although we have tried to represent all major biomes in our sample, we were unable to obtain informative material from wet regions of West Africa, where poor conditions for organic preservation militate against studies of culture-environment (cf. Nygaard and Talbot 1984).

Two matters of style deserve comment. Where differences exist between British and American orthography, we have accepted the author's choice of spelling. As regards radiocarbon dates, we have used the widely accepted convention of upper case letter designations (AD, BC) for calibrated dates and lower case designations (bp) for uncalibrated ones. We had intended to prepare this volume for publication within a year of the symposium, but various professional and personal commitments have conspired against us. Despite the delay we are pleased to see that the content of the papers presented here is by no means obsolete. But this should not lead us to be complacent. There is much yet to learn about human-environmental interactions during the later prehistory of Africa, and we hope that this volume will help identify useful avenues of investigation. —*John Bower and David Lubell*

References

Kryzyaniak, L. and M. Kobusiewicz (eds.). 1984. *Origin and Early Development of Food Producing Cultures in North-Eastern Africa*. Poznan: Polish Academy of Sciences and Poznan Archaeological Museum.

Lubell, D. 1984a. The Capsian palaeoeconomy in the Magrheb (abstract). In, Kryzyaniak and Kobusiewicz (eds.), pp. 453–455.

Lubell, D. 1984b. Paleoenvironments and Epi-Paleolithic economies in the Maghreb (*ca.* 20,000 to 5,000 B.P. In *From Hunters to Farmers: The Causes and Consequences of Food Production in Africa* (J. D. Clark and S. A. Brandt, eds.): pp. 41–56.

Lubell, D., A. Gautier, E. Leventhal, M. Thompson, H. Schwarcz and M. Skinner. 1982–83. The prehistoric cultural ecology of Capsian escargotieres. Part II: Report on investigations conducted during 1976 in the Bahiret Telidjene, Tebessa Wilaya, Algeria. *Libyca* 30–31:59–142.

Lubell, D., P. Sheppard and M. Jackes. 1984. Continuity in the Epipalaeolithic of northern Africa with an emphasis on the Maghreb. *Advances in World Archaeology* 3:143–191.

Nygaard, S. E. and M. R. Talbot. 1984. Stone age archaeology and environment on the southern Accra Plains, Ghana. *Norwegian Archaeological Review* 17:19–38.

Holocene Nile Floods and Their Implications for Origins of Egyptian Agriculture

By Fekri A. Hassan, Department of Anthropology, Washington State University, Pullman, Washington.

Introduction

The origin of agriculture in the Nile Valley has been one of the thorniest problems in Egyptian prehistory (Clark 1971, 1980). The present evidence suggests that agriculture emerged during the Middle Holocene, presumably by the middle of the seventh millennium bp. However, there is a break in the archaeological record from about 7140 to 6400 bp, and the Holocene sites predating this break do not seem to share too many technological similarities with the Neolithic/Predynastic sites. Although utilization of wild cereals in the Nile Valley dates back to the terminal Pleistocene, there is *no* evidence of domestic plants in Egypt prior to the Middle Holocene.

In my continuing search for a satisfactory explanation of the origins of agriculture, I emphasize, following the example set by Huzayyin (1939, 1950), Passarge (1940), and Butzer (1959), the importance of understanding the close link between the variations in Nile floods and the economic endeavors of the inhabitants of the Valley on the one hand, and on settlement location and preservation on the other. It is clear to me that we cannot begin to evaluate the nature of the existing archaeological record, not to mention explain agricultural origins, without an accurate reconstruction of the variations in Nile floods during the Holocene. In this chapter, I present my observations on the variations in Holocene Nile floods from my work on geoarchaeology in the western Fayum (Hassan, n.d.). I will attempt to integrate these observations with those made by Wendorf and Schild (1976), Butzer (1976, 1984), and others. I will then discuss the implications of the Nile record for the problem of agricultural origins in the Nile Valley.

The Holocene History of Nile Floods

Reconstruction of variations in Nile flood heights during the Holocene was initially attempted from the history of Nile aggradations under varying assumptions about the levels of Nile floods associated with aggradational phases. My studies (Hassan 1985a) lead me to think that the Nile tends to aggrade regardless of the height of the floods. It is during the transition from low Nile to high Nile episodes or vice versa that degradation prevails. This explains in part the short time associated with degradations compared with the longer time intervals associated with aggradations. For this reason, the utility of the depositional history of the Nile as a proxy to Nile flood heights is doubtful. Luckily, the geological record of Holocene Nile floods is not restricted to the succession of riverine deposits. In the Fayum Oasis, situated in a depression appended to the Nile Valley, a lake has existed during most of the Holocene period (Fig. 1.1), filled annually by an influx of Nile floodwater. Its level has thus varied with the changes in the height of Nile floods.

About 450 BC, Herodotus in a trip to the Fayum claimed that he saw a vast lake which he attributed to King Moeris. Accordingly, the historical lake is dubbed as Moeris Lake. In the early Ptolemaic period a land reclamation project cut off its natural connection with the Nile. Thereafter the level of the lake was progressively reduced from about 20 m above sea level to its present level at about 45 m below sea level in AD 1922. The modern remnant of Moeris Lake is Birket Qarun—a shallow brackish lake that covers about 200 km^2.

Geological investigations of the lake began around the turn of the century and led to the recognition of several ancient shorelines (Beadnell 1905; Brown 1892; Schweinfurth 1886). It was not until the 1930s after systematic investigations of these ancient shorelines by Thompson and Gardner (1934) and Sandford and Arkell (1929), that a detailed history of the lake began to take shape. In 1939, Ball evaluated existing evidence and presented a synthesis which was widely accepted until more recent investigations by the Combined Prehistoric Expedition (Said *et al.* 1972a, 1972b; Wendorf and Schild 1976) and by a Polish mission from the Jagellonian University (Kozlowski 1983). In 1981, I began a study of the Quaternary geology of western Fayum in collaboration with an American expedition led by Robert Wenke and Mary Allen Lane.

The results of investigations by the Combined Prehistoric Expedition place the earliest Holocene lake at 9000 bp. Several successive lakes are recognized and called Paleomoeris (12 m), Premoeris (15–19 m), Protomoeris (19–24 m), and Moeris lakes. The results suggest that the lake was rising during the Neolithic occupation and that it stood at about 23 m asl during the Old Kingdom.

My investigations in western Fayum confirm the presence of deposits belonging to two early Holocene lakes. The deposits are analogous to those associated with the Paleomoeris, Premoeris and Protomoeris lakes. Deposits of a middle Holocene lake (the Neolithic/Predynastic lake), and a late Holocene lake (historical lake Moeris) are also represented.

According to Wendorf and Schild (1976:222), the deposits of the Paleomoeris Lake were seen only in one locality and are represented by "fluvial" sand and diatomite. They estimate the age of this lake stage at about 9000 bp. They note deep fossil desiccation cracks in the top of the upper diatomite unit. The Polish mission located "lake chalk and diatomites" dated at one site to 7440 ± 60 bp (Bln-Z336) at about 13 m above sea level. They also noted drying of the lake and precipitation of anhydrites.

In western Fayum, I observed a sequence of diatomaceous deposits with fossil desiccation cracks at about 11–12 m asl. The diatomaceous deposits are associated with a lake beach deposit. A Qarunian site was found on the surface and interspersed with the top layer of the beach facies. The dates associated with this site (Wenke, personal communication) are:

FS-2	TS-8	Level 2	8220 ± 105 bp	Beta-4871	Charcoal
FS-2	TS-12	Level 4	7720 ± 70 bp	Beta-4872	Charcoal
FS-2			7600 ± 70 bp	Beta-4180	Charcoal

The weighted average of these dates is 7715 ± 45 bp, using Long and Rippeteau's (1974) method.

These diatomaceous deposits are thus similar in age to those reported by the Polish mission and younger than those previously noted by Wendorf and Schild (1976). At Site E29G1, Area F, Wendorf and Schild (1976) report sand with intercalations of dark, grayish brown, loose, powdery sediments overlying the desiccated surface of the Paleomoeris diatomaceous deposits. A date of 8100 ± 130 bp (I-4128) on charcoal was obtained at that site from an occupation at 15 m asl contemporaneous with a mat of swampy deposit. At Site E29H1, a date of 8070 ± 115 bp (I-4126) was obtained from a similar layer (Wendorf and Schild 1976:192). Another date of 7500 ± 125 bp was obtained from a terminal Paleolithic occupation in a dark layer at Site E29G3 at 12.2 m asl. Wendorf and Schild assign these deposits to the Premoeris stage and note that the stage was followed by a drop of the lake to below 12 m asl. After this recession the lake rose again to 19 m asl and possibly 24 m asl forming the Protomoeris Lake. The deposits of this lake are represented by sand and sandy silt with intercalations of swampy deposits (Wendorf and Schild 1976:201). The deposits are similar to those of the Premoeris lake, but an unconformity separates the deposits of the two lakes. One date on burnt shell is associated with the deposits of the Protomoeris lake. The date is 7140 ± bp (I-4129).

It thus appears that the sequence consists of Paleomoeris diatomaceous deposits predating 8100 bp, followed by a Premoeris Lake well dated to 8100 bp. The recession of the Premoeris Lake was associated with the formation of diatomaceous deposits similar to those of the Paleomoeris low lake. This lake stage between the Premoeris and

Holocene Nile Floods

Protmoeris lakes is well dated to about 7800 bp. The date of 7440 ± 60 bp from the top of the diatomaceous deposits, and the date of 7500 ± 125 bp on swampy deposits, suggests that the transition from the diatomaceous lake to the swampy conditions of the higher Protomoeris lake occurred around 7450 bp. The Protomoeris lake, given the 7140 ± 120 bp date, was followed by a recession preceding the rise of the Neolithic lake. Qarunian sites (MOE2, MOE2b, MOE2c) at 2 m below sea level (Mussi *et al.* 1980:Fig. 1a) probably belong to this phase of lake recession.

The deposits of the Neolithic lake in northern Fayum are recorded at Site Kom W (E29H2) and consist mostly of pale brown sand (Wendorf and Schild 1976:213). The dates associated with the Neolithic occupation are 5810 ± 115 bp (I-4172) from a level at 17 m asl, and 5860 ± 115 bp (I-4131) at 15 m asl. In western Fayum, Neolithic lake deposits are encountered about 2.5 km southwest of Madient Quta and consist mostly of pale brown sand. A Neolithic beach facies was also encountered in association with the Neolithic Site FS-1 (Wenke *et al.* 1983). The Neolithic archaeological remains are widely dispersed at the surface. Hearth stones, fish bones, and artifacts were located in the sand of the lake. The sand is medium grained, very pale brown (10YR 7/4). Beach deposits (Section Qq54) of the high stand of the lake overlie the artifact-bearing sand and consist of very coarse-grained to medium-grained sand with flat, well-rounded shingle. In another section (Qq55), deposits of a swampy facies were exposed. Lithic artifacts and ceramics were found embedded in very fine-grained sand underlying a dark, carbonaceous silt at another locality (GS2). Site FS-1 is dated to 5160 ± 70 bp (Beta-4181) on charcoal (Wenke personal communication). The lake deposits are at about 20 m asl.

The results of the Polish mission to northern Fayum suggest a separation between a "white sand" of Neolithic age and a "brown sand" of Dynastic age. The results from the Combined Prehistoric Expedition and my own observations indicate that the Neolithic sediments are predominantly pale brown (from pale yellow to brown) sand with intercalation of dark carbonaceous layers and white sand. The two types of sand are most probably related to differences in facies rather than age. The dates obtained by the Polish mission (Kozlowski 1983:119) indicate that the top of Neolithic lake deposits situated at 20 m asl at Site QS V/79 is dated to 6075 ± 50 bp (Bln-2335) and 5990 ± 60 bp (Gd-693). The early dates now available on the Fayum Neolithic indicate that the Neolithic lake probably rose before the end of the seventh millenium bp and may have crested as early as 6500 bp. A list of dates for the Fayum Neolithic and the late Fayum Neolithic (also referred to as the Predynastic) is given in Table 1.1 in radiocarbon years before present.

Table 1.1: Dates, in radiocarbon years before present, for Fayum Neolithic and Late Fayum Neolithic (Predynastic) sites

Site		Date bp	Lab No.	Reference
Fayum Neolithic				
I/79		5540 ± 70	Od-1140	Ginter et al. 1982
1/79		5555 ± 60	Bln-2333	Ginter et al. 1982
1/79		5645 ± 55	Bln-2334	Ginter et al. 1982
VIE/81		5650 ± 70	Gd-1495	Ginter personal communication
Kom K		5810 ± 100	1-4172	Wendorf and Schild 1976
Site R		5860 ± 115	1-4131	Wendorf and Schild 1976
V/79		5990 ± 60	Gd-693	Ginter et al. 1982
V/79		6075 ± 50	B1n-2325	Ginter et al. 1982
Upper K		6095 ± 250	C-457	Libby 1955
X/81-5		6290 ± 110	Gd-980	Ginter personal communication
X/81-2		6290 ± 100	Gd-979	Ginter personal communication
Upper K		6391 ± 180	C-550 & 551	Libby 1955
IX/81		6380 ± 60	Gd-1499	Ginter personal communication
X1/81		6480 ± 70	Od-2021	Ginter personal communication
Late Fayum Neolithic (also referred to as the Predynastic)				
VIIA/80		5070 ± 110	Gd-895	Ginter et al. 1982
VIIA/80		5480 ± 110	Gd-977	Ginter personal communication
VIIA/80		5160 ± 120	Gd-915	Ginter et al. 1982
	Average	5224 ± 90		
FS-I		5160 ± 70	Beta-4181	Wenke personal communication
FS-2		4960 ± 160	Beta-4182	Wenke personal communication
FS-2		5475 ± 70	Beta-4874	Wenke personal communication
	Average	5284 ± 120		
X-81		5330 ± 100	Gd-978	Ginter personal communication
N. Fayum		5388 ± 45	BM-530	Barker et al. 1971
VID/80		5410 ± 110	Gd-903	Ginter et al. 1982

Sites and artifacts associated with the Late Neolithic (Predynastic) indicate that the lake was still high by about 5160 bp as indicated by the remains at Site FS-l (Wenke *et al.* 1983). Afterwards the lake began to shrink. Gardner (in Caton-Thompson and Gardner 1934:15) recognized shorelines associated with the shrinking Neolithic lake at 10 m and 4 m asl.

In western Fayum, I located diatomaceous deposits at 15 m below sea level followed by progradation to about 8 m below sea level in a drain at Ezbet Gebel Saad. The sequence consists of 1.7 m of diatomaceous earth overlaid by 3.5 m of silt, which are in turn overlaid by 1 m of clayey silt capped by 4 m of silt. The sequence represents a cycle beginning with the shallow lake deposits with gradual change into deeper basin conditions. The stratigraphic position of these deposits strongly suggests that they belong to a lake stage predating the Old Kingdom high stand and postdating the Neolithic lake. The shrinking of the Neolithic lake thus seems to have reached a nadir sometime during the Early Dynastic (Archaic) period. Butzer (1976:28), in a discussion of the results obtained by Bell (1970), suggests that the record of 63 annual floods from the early dynasties indicates that flood levels show a general decline that was most rapid during the first and early second dynasty (about 3000 BC or 4350 bp). The year 14 of Ninetier (about 2913 BC) experienced one of the lowest floods (Butzer 1984).

The evidence of lake level during the Old Kingdom suggests that after its drastic recession the lake rapidly rose to about 24 m asl by 3890 ± 45 bp (about 2460 BC/Fifth Dynasty). The average lake level was around 18–22 m during the Old Kingdom (Butzer 1976:36). The rapid rise of the Old Kingdom lake seems to have happened after the Fourth Dynasty (2650–2500 BC or 4050–3920 bp) when the lake level stood at about 2 m below sea level (Caton-Thompson and Gardner 1934:15). The Old Kingdom artifacts noted by Caton-Thompson and Gardner (1929:46) below Qasr el-Sagha Temple at 4 and 13 m asl probably belong to the rising level of the Old Kingdom lake from its nadir during the early dynasties to its maximum after the Fourth Dynasty.

There are no dated deposits dating to the First Intermediate, but it is generally believed that this was a period of low Nile floods. Bell (1971, 1975) examined numerous texts from that period and concluded that several catastrophically low Nile floods occurred between 2250 and perhaps 1950 BC (3730–?3500 bp). Butzer (1984:107) also concluded from an evaluation of the historical documents that at least two episodes of catastrophic Nile failures occurred around 2200 BC and 2002 BC (about 3700 and 3540 bp).

Lake level during the Middle Kingdom was relatively high. According to Ball (1939:202–7) the lake level stood at about 16.5 m during the low Nile season at the time of Amenmhat 1 (1991–1961 BC or 3535-3510 bp.). Butzer (1976:37) suggested that the lake stood at about 15 m asl with a level of about 18 m during the flood season. Butzer (1984:107) also notes that Nile flood records indicate that floods higher than at present occurred from 1840 to 1770 BC (about 3400 to 3365 bp). But he also notes (ibid.) that low floods occurred at about 1200 BC (about 2925). The lake level may thus have been low at that time.

Around 450 BC, Herodotus visited the Fayum and reported a vast lake. Ball suggested that the lake oscillated seasonally between 17.8 and 20 m at that time. In 1982, I located lake shoreface sediments underlying a fallen block from the pedestals of the Biyahmu colossi witnessed by Herodotus (he thought looking from a distance that they were pyramids). The sediments were above the base on which the pedestals were erected and indicate that the lake must have risen to about 20 m following the construction of the colossi suggesting that the lake may have indeed stood at about 20m asl at 450 BC (2325 bp) when Herodotus was in the Fayum.

In early Ptolemaic times a reclamation project controlled free influx of Nile floods into the lake. By 280 BC the lake stood less than 5 m below sea level. During the second century AD it stood less than 7 m below sea level and dropped to 15 m below sea level during the next century (Ball 1939:210; Sahafei 1960). By AD 1245 the lake dropped to 30 m below sea level (Shafei 1960:194, 1940:32), and to its present level at about 45 m below sea level at AD 1922 (Shafei 1960:194).

A reconstruction of Holocene variations in the levels of Lake Qarun from geological observations in combination with historical data is shown on Fig. 1.2.

The Middle Holocene Hiatus

The early and middle Holocene industries in the Nile Valley consist of the Arkinian, Shamarkian, Kabian, and Qarunian (Hassan 1980).

The Qarunian, as noted above is known from the Fayum Oasis. The sites date from about 8220 bp to 7180 bp (Wendorf and Schild 1976; Wenke *et al.* 1983). Abundant fish remains suggest that fishing was an important subsistence activity. Bones of wild cattle and hartebeest and microlithic geometrics indicate the practice of hunting. Grinding stones were present and though we may assume that they were used for processing plants, there is no evidence for domestication.

At Hierakonpolis, where Predynastic sites (Amratian to Gerzean) are located (Hoffman 1982), the earlier sites in the region are those of the Kabian industry (Vermeersch 1978). The Kabian, dating around 8000 bp, did not provide any evidence for domestication.

Farther south in Nubia, the early Holocene industries include the Arkinian and the Shamarkian. The Arkinian (Schild *et al.* 1968) is known from the Wadi Halfa area and dates to 10,700 bp. Fish and hartebeest bones and grinding stones are found at the sites. The Shamarkian (Schild *et al.* 1968) dates to 8900 bp. Fish bones (*Lates*) were recovered from one site, but other animal bones are rare to absent.

The radiocarbon dates from these pre-Neolithic Holocene sites (all on charcoal) in radiocarbon years bp are given in Table 1.2.

The earliest dated "Neolithic" settlements in the Nile Valley, Egypt, are in the Fayum depression. The oldest of these settlements is well dated to 6350 ± 50 bp or 5230 ± 50 BC (Hassan 1985b) based on dates from three sites (IX, X, XI) at Qasr el Sagha (Ginter *et al.* 1982). At Merimda Beni Salama, the dates from early levels indicate that the settlement was established beginning about 5900 bp or 4800 BC (Hassan 1985b) based on several dates by Olsson (1959) and more recent determinations kindly

Holocene Nile Floods

provided to me by R. Kuper from the recent excavations by the German mission. In Upper Egypt, agricultural communities from Hemamieh are dated to about 5380 bp or 4250 BC (Hassan 1984c). The oldest industry with ceramics in Upper Egypt is at El-Tarif near Luxor (Ginter *et al.* 1982) and is dated to 6310 ± 80 bp or 5185 ± 120 BC. Later sites from the Nagada I phase date to about 5000 bp or 3800 BC (Hassan 1984b). In the Sudan, the earliest "Neolithic" communities are dated to about 5900 bp or 4800 BC at Umm Direiwa I (Haaland 1979) and Islang (El-Anwar 1981).

There is thus clearly a hiatus in the archaeological record between the Holocene pre-Neolithic sites and the Neolithic and Predynastic sites. The majority of the pre-Neolithic Holocene sites date no later than 7400 bp; the youngest date is 7140 bp. The oldest date on the Neolithic/Predynastic sites or Post-Paleolithic sites with pottery in the Nile Valley is 6350 bp. There is thus a hiatus of at least 800 years in the record of radiocarbon dates.

Table 1.2: Radiocarbon dates on charcoal in radiocarbon years bp for pre-Neolithic Holocene sites

Industry/Site	Date bp	Lab. No.	Reference
Arkinian			
DIW-I	10,670 ± 110	SMU-581	Wendorf et. al 1979
Shamarkian			
DIW-51	8860 ± 90	SMU-582	Wendorf et. al 1979
Kabian			
ElKab 1	8340 ± 160	Lv-393	Vermeersch 1978
	7990 ± 150	Lv-464	Vermeersch 1978
	7930 ± 160	Lv-465	Vermeersch 1978
ElKab 2	8095 ± 75	GrN-7188C	Vermeersch 1978
	7885 ± 50	GrN-7190	Vermeersch 1978
ElKab 3	8090 ± 75	GrN-7189C	Vermeersch 1978
Qarunian			
FS-2	8220 ± 105	Beta-4871	Wenke personal communication
FS-2	7720 ± 70	Beta-4872	Wenke personal communication
FS-2	7600 ± 70	Beta-4180	Wenke personal communication
E29GIF	8100 ± 130	1-4128	Wendorf and Schild 1976
E29GIE	7140 ± 120	I-4129	Wendorf and Schild 1976
EI9HI	8070 ± 115	1-4126	Wendorf and Schild 1976
E29G3	7500 ± 125	1-4130	Wendorf and Schild 1976
I/79	8835 ± 890	Gd-709	Kozlowski 1983
II/79	7440 ± 60	Bln-2336	Kozlowski 1983

This hiatus is most likely a result of the drop in the Nile level at about 7000 bp and perhaps a century earlier, signaling a major recessional episode that may have lasted until 6500 bp. The record from the Fayum Oasis discussed above shows a drop of the level of the lake following the Protomoeris stage (7140 bp). Given a date for a high Nile level at 7060 ± 120 bp (Y-1664) from Nubia (Wendt 1966), the main drop probably began about 7000 bp. The drop was sufficiently long to allow a reddish paleosol to develop on the lacustrine deposits of earlier lake stages that were exposed under subaerial conditions. The paleosol was first noted by Wendorf and Schild (1976:225). I recorded a paleosol in a similar stratigraphic position post-dating the Qarunian site at FS-2 in western Fayum. The drop of the lake was so drastic that Wendorf and Schild (1976:225) suspect that the lake may have totally dried up from 7000 to 6000 bp!

Desert Origins

It should be noted here that an interval of severe aridity occurred in the Western Desert of Egypt at about 7100–6900/6500 (Hassan 1984c). The synchronism between the low Nile floods and dry conditions in the desert is not surprising since the source of rain in the Western Desert is supposed to be an extension of the monsoonal rain belt from which the Nile is also fed (Wendorf and Schild 1980). Accordingly, I propose that sites from about 7000 to 6500 bp were placed close to the channel of the Nile in response to the weak Nile floods. These sites are likely to have been destroyed by the higher floods of the Nile during the interval from 6500 to 5000 bp when the Neolithic lake in the Fayum was expanding. The high Nile floods most likely led people to place their settlements in the desert bordering the flood plain and on the low desert terraces overlooking the flood plain where all of the known Predynastic sites are found as at Hemamih, Nagada, and Hierakonpolis. The drop in the height of Nile floods by about 5000 bp, synchronous with the transition from Nagada I to Nagada II in Upper Egypt, was associated most likely with greater emphasis on agriculture and led to the relocation of settlements closer to the channel. This is clear from the distribution of the ceramics in the Hierakonpolis area (M. Hoffman, personal communication), and the lack of Gerzean villages and hamlets in the Nagada region (the only settlements are the urban ones at South Town and North Town). This also explains the extreme rarity of terminal Predynastic (Nagada III) sites in Egypt immediately preceding the nadir of Nile recession during the early dynasties.

There is at present no evidence for either plant or animal domestication in the Nile Valley prior to the later part of the seventh millennium bp. Although it may be suspected that agriculture might have been developed independently in the Nile Valley during the interval of the hiatus, the present sample of early Holocene sites suggests that the known sites from 10,670 to 7140 bp do not have any evidence for domestication of either animals or plants. The subsistence regime is likely to have been one of fishing, hunting, and plant gathering. In settlements, such as those at which the Qarunian has been found, fishing might have been the primary subsistence activity. Numerous hearth areas suggest that fish were probably dried and stored.

There is thus sufficient evidence to preclude assuming that agriculture or food raising was independently developed in the Nile Valley prior to its inception in the eastern Sahara. Although future evidence may reveal otherwise (Clark 1980), the present data on desertification during the early part of the seventh millennium bp, and its probable impact on the desert Neolithic pastoralists in a manner analogous to that affecting the Sahel people today, strongly suggest that food production was introduced into the Nile Valley from the surrounding desert regions. This is also suggested by similarities between some Neolithic Saharan and Nilotic elements (Hassan and Holmes 1985). In addition, the known sites from the tenth millennium bp, in the period predating the appearance of agriculture, are in general different in their technological and typological traditions from the Neolithic/Predynastic agricultural sites. There is also no evidence of pottery from the pre-agricultural sites, except at El-Tarif.

The middle Holocene dry period marking the beginning of desertification occurred at about 7100–6900/6600 bp (cf. Hassan 1984d; Haynes, 1982c; Wendorf and Hassan 1980). At that time, at least some of the inhabitants of the Western Desert of Egypt practised herding of cattle and ovicaprids, as well as cultivation of barley (Clark 1980:567; A. B. Smith 1980, 1984). Domestication, "management," or "taming" of cattle dating to 9000 bp or even earlier has been suggested (Close 1984; Wendorf and Schild 1980:266). Cultivation of barley is dated at Nabta to 8100 bp (Stemler and Falk 1980).

The food producing communities of the Western Desert were most likely nomadic and not unlike present pastoral nomads in the West African Sahel and the Hoggar (Clark 1980:567; S. E. Smith 1980). This mode of life was, as it is today, necessitated by the great variability of rainfall from one year to the next and the unpredictability of where good pastures and water are likely to be found.

It was the impact of the mid-Holocene droughts that are likely to have led to the dispersal of pastoral Saharan folk, already in possession of rudimentary agriculture and pottery into the Nile Valley and probably south of the Sahel (Hassan 1977, 1984a, 1984b).

Acknowledgements

I thank Robert Wenke and Mary Allan Lane for the opportunity to contribute to the Fayyum Project, the Geological Survey of Egypt for logistic support in the field, Jill Kaplan for field assistance, and T. El-Diftar and D. L. Holmes for participation in the excursion to Biyahmu.

References

Ball, J. 1939. *Contributions to the Geography of Egypt*. Cairo: Egypt Government Press.

Barker, H., Burleigh, R. and Meeks, N. 1971. British Museum natural radiocarbon measurements Vll. *Radiocarbon* 13:166.

Beadnell, H. J. L. 1905. *The Topography and Geology of the Fayum Province of Egypt*. Cairo: Egypt Survey Department.

Bell, B. 1970. The oldest records of the Nile floods. *Geogr. Journal* 136:569–73.

———. 1971. The dark ages in ancient history. *Amer. J. Archaeol.* 75:1–26.

———. 1975. Climate and history of Egypt: the Middle Kingdom. *Amer. J. Archaeol.* 79:223–69.

Brown, R. 1892. *The Fayum and Lake Moeris*. London: Edward Sandford.

Butzer, K. W. 1959. Environment and human ecology in Egypt during predynastic and early dynastic times. *Bulletin de la Société Géographie d'Égypte* 32:43–87.

———. 1976. *Early Hydraulic Civilization in Egypt*. Chicago: The University of Chicago Press.

———. 1984. Long-term Nile flood variation and political discontinuities in Pharaonic Egypt. In *From Hunters to Farmers* (eds J. D. Clark and S. A. Brandt): pp. 102–12. Berkeley: University of California Press.

Caton-Thompson, G. and Gardner, E. 1929. Recent work on the problem of Lake Moeris. *Geogr. Journal* 73:20–60.

_____. 1934. *The Desert Fayum*. London: Royal Anthropological Institute.

Clark, J. D. 1971. A re-examination of the evidence for agricultural origins in the Nile Valley. *Proceedings of the Prehistoric Society* 37:34–79.

_____. 1980. Human populations and cultural adaptations in the Sahara and Nile during prehistoric times. In *The Sahara and the Nile* (eds M. A. J. Williams and H. Faure): pp. 527–82. Rotterdam: Balkema.

Close, A. E. 1984. Current research and recent radiocarbon dates from northern Africa II. *Journal of African History* 25:1–24.

El-Anwar, S. 1981. Archaeological excavations on the west bank of the River Nile in the Khartoum area. *Nyame Akuma* 18:42–5.

Gardner, E. W. 1929. The origin of the Fayum Depression: a critical commentary on a new view of its origin. *Geogr. Journal* 74:371–83.

Ginter, V. B., Kozlowski, J. K., Pawlikowski, M. and Silwa, J. 1982. El-Tarif und Qasr el-Sagha, Forschungen zur Siedlungsgeschichte des Neolithikums, der Frühpredynastischen Epoche und des Mittleren Reiches. *Mitteilungen des Deutschen Archäologischen Instituts Abteilung Kairo* 38:97–129.

Haaland, R. 1978. Report on the 1979 season in the Sudan. *Nyame Akuma* 14:62.

Hassan, F. A. 1977. Holocene palaeoclimate in North Africa. Paper read at the 4th meeting of the Society of Africanist Archaeologists in America, New Orleans.

_____. 1980. Prehistoric settlements along the main Nile. In *The Sahara and the Nile.* (eds M.A. J. Williams and H. Faure): pp. 421–50. Rotterdam: Balkema.

_____. 1984a. Toward a model of agricultural development in Predynastic Egypt. In *Proceedings of the lst Symposium on the Late Prehistory of the Nile Basin and the Sahara* (eds L. Krzyzaniak and M. Kobusiewicz). Poznan.

_____. 1984b. Radiocarbon chronology of Predynastic Nagada settlements, Upper Egypt. *Current Anthropology* 25:681–83.

_____. 1984c. A radiocarbon date from Hemamieh, Upper Egypt. *Nyame Akuma* 24/25.

———. 1984d. Mid-Holocene desertification and human responses in the Western Desert of Egypt. Paper presented at the 17th Annual Chacmool Conference, Calgary.

———. 1985a. Fluvial systems and geoarcheology in arid lands: with examples from North Africa, the Near East, and the American Southwest. In *Archaeological Sediments in Context* (eds J. K. Stein and W.R. Farrand): pp. 53–68. Orono: Center for the Study of Early Man.

———. 1985b. Radiocarbon chronology of Neolithic and Predynastic sites in Upper Egypt and the Delta. *The African Archaeological Review* 3:95-111.

———. n.d. Quartenary geology of western Fayum, Egypt: A contribution to the Fayyum Archaeological Project. Ms.

Hassan, F. A. and Holmes, D. L. 1985. The archaeology of the Umm el-Dabadib area, Kharga Oasis, Egypt. *FRSU Research Project Report 8Z035.* Cairo: Cairo University.

Haynes, C. V. 1982. Quartenary geochronology of the Western Desert. In *First Them. Conf. Remote Sensing of Arid and Semi-Arid Lands.*: pp. 297–311. Cairo.

Hoffman, M. 1982. *The Predynastic of Hierakonpolis—An Interim Report.* Egyptian Studies Association Pub. No. 1, Cairo University Herbarium. Oxford: The Alden Press.

Huzayyin, S. A. 1939. Some new light on the beginnings of Egyptian civilization. *Bulletin de la Société Royale de Géographie d'Égypte* 20:203–73.

———. 1950. Origins of Neolithic and settled life in Egypt. *Bulletin de la Société Royale de Géographie d'Égypte* 23:175–81.

Kozlowski, J. K. 1983. *Qasr el-Sagha 1890.* Warszawa-Krakow: Panstwowe Wydawnictwo Naukowe.

Libby, W. F. 1955. *Radiocarbon Dates.* Chicago: University of Chicago Press.

Long, A. and Rippeteau, R. 1974. Testing contemporaneity and averaging radiocarbon dates. *American Antiquity* 39:205–15.

Mussi, M., Caneva, I. and Zarattini, A. 1980. More on the terminal Palaeolithic of the Fayyum Depression. Paper presented at the *International Symposium on the Origin and Early Development of Food-Producing Cultures in North-Eastern Africa*, Poznan.

Olsson, I. 1959. Uppsala natural radiocarbon measurements I. *Radiocarbon* 1:87–102.

Passarge, S. 1940. Die urlandschaft Ägyptens und die localisierung der Wiege der Altäggyptischen Kultur. *Nova Acta Leopolding* 9:77–152.

Said, R., Albritton, C., Wendorf, F., Schild, R. and Kobusiewicz, M. 1972a. Remarks on the Holocene geology and archaeology of Northern Fayum Desert. *Archaeologia Polona* 13:7–22.

_____. 1972b. A preliminary report on the Holocene geology and archaeology of the northern Fayum Desert. In *Playa Lake Symposium* (ed C. C. Reeves, Jr.): pp. 41–61. Lubbock: Icassals publication no. 4.

Sandford, K. S. and Arkell, W.J. 1929. *Paleolithic Man and the Nile-Faiyum Divide*. Chicago: University of Chicago Press. Oriental Institute Publication 10.

Schild, R, Chmielewski, M. and Wieckowska, H. 1968. The Arkinian and Shamarkian Industries. In *The Prehistory of Nubia, Vol. 2* (ed F. Wendorf): pp. 651–767. Dallas: Southern Methodist University Press.

Schweinfurth, G. 1886. Reise in das Depressions-gebiet im Umkreise des Fajum im Januar 1886. *Z. Ges. Erdk.* 21:96–149.

Shafei, A. 1940. Fayoum irrigation as described by Nabulsi. *Bulletin de la Société Royale de Géographie d'Égypte* 20:283–327.

_____. 1960. Lake Moeris and Lahun Mi-Wer and Ro-Hun: the great Nile control project executed by the Ancient Egyptians. *Bulletin de la Société Royale de Géographie d'Égypte* 33:187–215.

Smith, A. B. 1980. Domesticated cattle in the Sahara and their introduction into West Africa. In *The Sahara and the Nile* (eds H. A. J. Williams and H. Faure): pp. 489–501. Rotterdam: Balkema.

_____. 1984. The origins of food production in Northeast Africa. *Palaeocology of Africa* 16:317–24.

Smith, S. E. 1980. The environmental adaptation of nomads in West American Sahel: a key to understanding prehistoric pastoralists. In *The Sahara and the Nile* (eds M. A. J. Williams and H. Faure): pp. 467–87. Rotterdam: Balkema.

Stemler, A. and Falk. R. 1980. A scanning electron microscope study of cereal grains from Nabta Playa. In *Prehistory of the Eastern*

Sahara (eds F. Wendorf and R. Schild): pp. 393–399. New York: Academic Press.

Vermeersch, P.M. 1978. *Elkab II, L'ElKabien, Epipaléolithique de la Vallée du Nil Egyptien*. Bruxelles: Fondation Egyptologique Reine Elisabeth.

Wendorf, F. and Hassan, F. 1980. Holocene ecology and prehistory in the Egyptian Sahara. In *The Sahara and the Nile* (eds M. A. J. Williams and H. Faure): pp. 407–19. Rotterdam: Balkema.

Wendorf, F. and Schild, R. 1976. *Prehistory of the Nile Valley*. New York: Academic Press.

_____. 1980. *The Prehistory of the Eastern Sahara*. New York: Academic Press.

Wendorf, F., Schild, R., Close, and Haas, H. 1979. A new radiocarbon chronology for prehistoric sites in Nubia. *Journal of Field Archaeology* 6:219-23.

Wendt, W. E. 1966. Two prehistoric archaeological sites in Egyptian Nubia. *Postilla* 102:1–46.

Wenke, R. J., Buck, P., Hanley, J. R., Lane, M. E., Long, J. and Redding, R. R. 1983. The Fayyum archaeological project: preliminary report of the 1981 season. *American Research Center in Egypt Newsletter* 122:25–40.

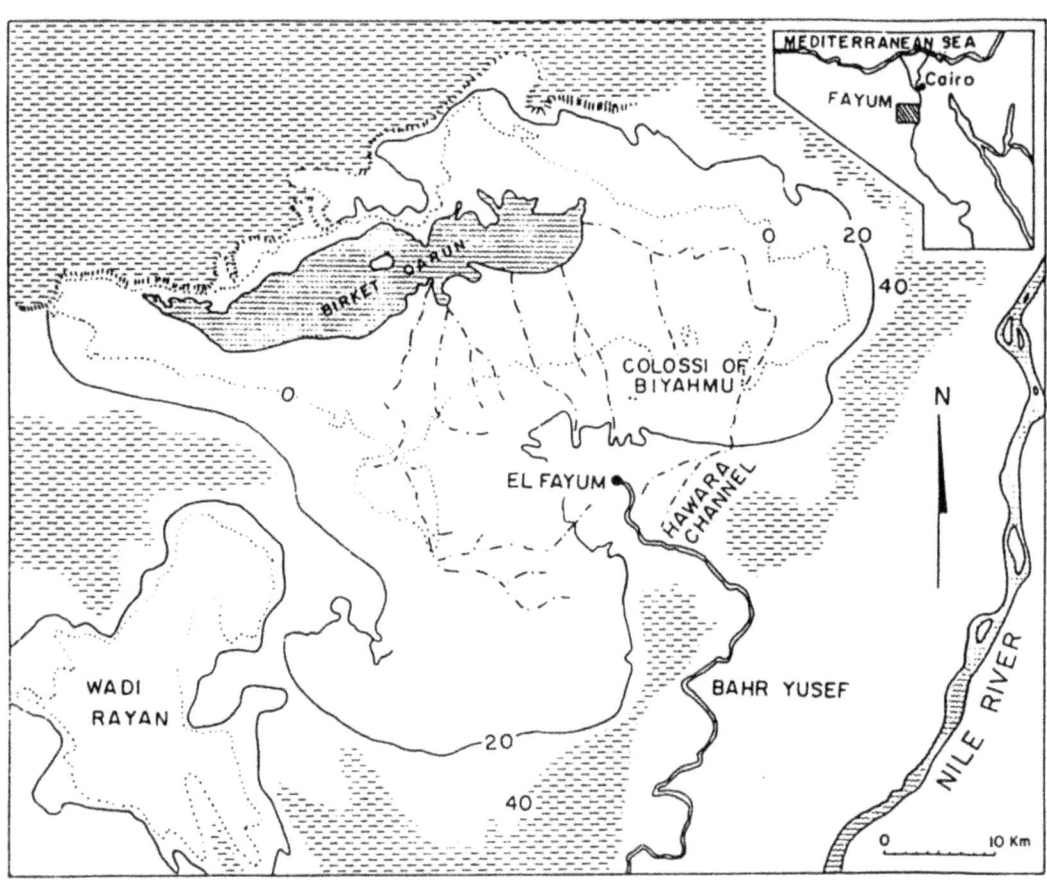

Fig. 1.1: Location map of Birket Qarun, the Fayum, Egypt.

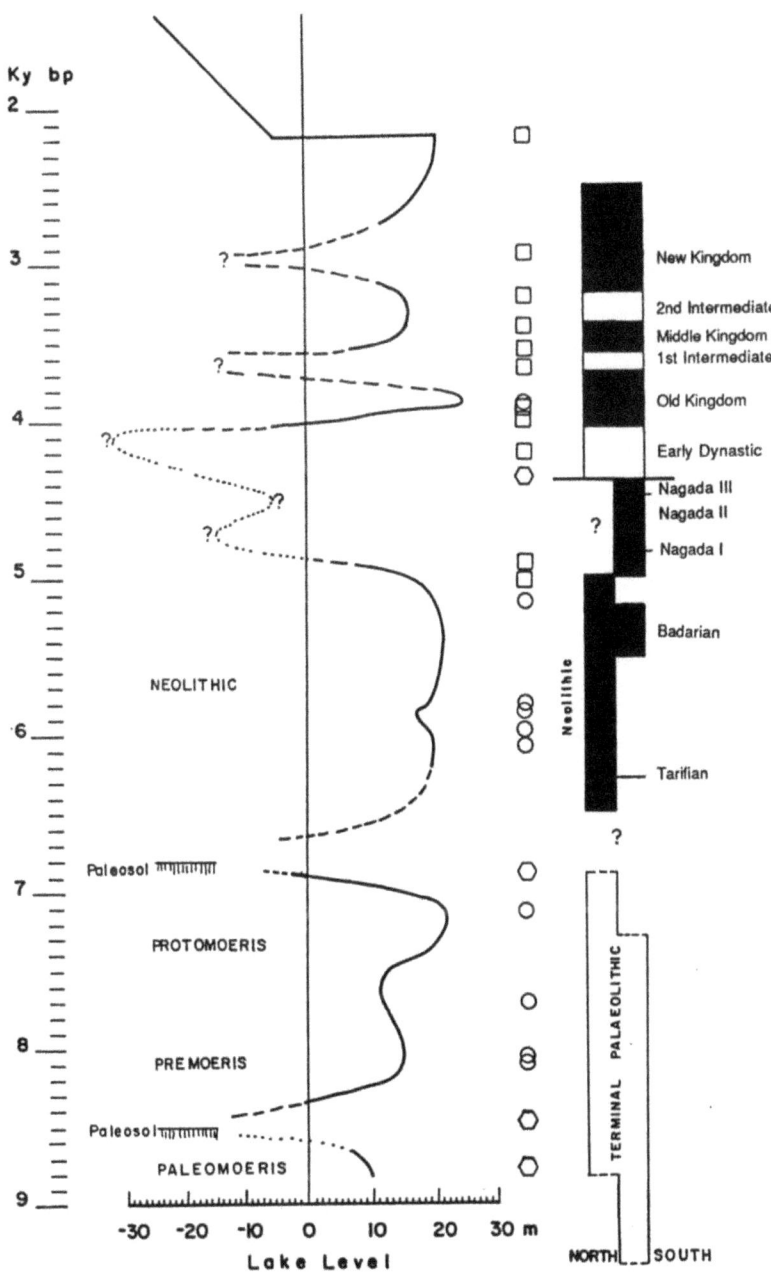

Fig. 1.2: Chart showing variation in lake level of the historical Moeris lake and its ancestral prehistoric lakes. The curve is reconstructed from elevation of lake deposits relative to sea level, historical accounts of lake levels and Nile floods. Circles show radiocarbon dates on lake deposits; hexagons show stratigraphic age estimates; and rectanges, historical dates. The radiocarbon dated chronology of terminal Paleolithic and Neolithic/Predynastic sites from northern and southern Egypt is also shown. The chronology of dynastic Egypt is based on historical dates.

Holocene Nile Floods

Climatic Change and Man in the Sahara

By Nicole Petit-Maire, Laboratoire de Géologie du Quaternaire, CNRS, Marseille, France.

Introduction

Man being very sensitive to environmental variations, the equilibrium of our societies is particularly fragile in those areas which suffer frequent and severe alterations of their biotope; the Saharo-Sahelian belts belong to these unstable zones due to the high variability of monsoon or subtropical precipitation (Rognon 1976; Maley 1983). In the last ten years, an increasing number of observations has allowed the correlation of their climatic evolution with the global changes recorded in oceanic cores, as well as with the dynamics of human populations and cultures.

This chapter will provide some data from research in the Fezzan, western Sahara, and the Taoudenni Basin, that have been collected since the synthesis by Desmond Clark (1980).

The Climatic Optimum at *ca.* 125,000 bp and the End of the Acheulian

The rapid rise of sea level (Eemian) and temperatures which began at *ca.* 140,000 bp, culminated at *ca.* 125,000 bp (Shackleton 1981, 1982; Lorius *et al.* 1985) with a marked increase in insolation (Berger 1979, 1981, 1984). In the central Sahara, this climatic optimum coincided with a major lacustrine development. To the north of the Murzuk sandsea, in a region where today the annual rainfall amounts to only 30 mm, a lake fed by aquifer rise and local runoff extended over 2000 km^2 at 27°30'N. This lake existed continually for more than 100,000 years (Gaven *et al.* 1980, 1982:Fig. 1) with salinities varying from 3 to 10°/oo and depths varying from 40 to 50 m (Petit-Maire 1979; Petit-Maire *et al.* 1980, 1982).

Surface archaeological sites are found throughout this area in Mali (Aumassip and Petit-Maire 1982): terminal Acheulian on the substructural flats above the valley; small (<8 cm) piriform bifaces, mixed with terminal Acheulian and Levallois flakes or blades along the upper shorelines of the lake; exclusively Levallois assemblages at the foot of the thick lumachelle mounds indicating a molluscan thanatocoenosis due to lake regression (Plate 2.I); Aterian and Neolithic unassociated with the shores.

According to their location relative to the lake deposits, the Levallois sites may be situated in time as a little younger than the lake main regressive phase, which is consistent with the age suggested by McBurney (1967) for the earliest Levallois- Mousterian in Cyrenaica, *ca.* -80,000.

Climatic Change and Man in the Sahara

We leave to archaeologists the interpretation of these data. However, one can underline the correlation of an important humid phase in the Sahara during the Mid-Pleistocene interglacial with the end of the old Acheulian civilisation. The occurrence and long duration of a hospitable biotope probably allowed sedentary life and invention, as well as easier contacts and transmission of new techniques.

Upper Pleistocene Climates and the Aterian

The progressive deterioration of climate recorded after 115,000 bp by isotopic curves was not regular. In the Sahara, repeated fluctuations between environmental degradation and regeneration are recorded. They differ widely with latitude (Durand *et al.* 1983) and imply either heavy stress or new territorial possibilities for prehistoric groups with probable response in terms of migrations and cultural change. The climatic and archaeological sequence of the Sahara between -100,000 and -50,000 is still little known. An arid phase occurred in southern Chad between -50,000 and -40,000 (Durand *et al.* 1983). It was followed between -40,000 and -20,000 by one or two humid episodes that have been observed right across the continent, from the Red Sea to the Atlantic, and which are associated with the Aterian (Alimen 1966; Butzer 1980; Chamard 1973; Conrad 1969; Durand *et al.* 1983; Faure 1962; Gasse 1975; Rognon 1976; Servant 1973; Servant and Servant-Vildary 1980; Tillet 1983; Williams and Adamson 1980).

In the Fezzan, the large Shati lake became a saline sebkha, but episodes of fluvial activity or small lakes are dated at - 40,000 and 26,000 to 22,500 bp on freshwater molluscs (Petit- Maire 1986). Aterian pedunculate points occur on the banks of the wadis or channels (Aumassip and Petit-Maire 1982).

In northern Mali, vast extensions of lacustrine limestones or large clayey pans, both with *Melania tuberculata*, are found in flat areas between 19°N and 22°30'N. At Kesret-el-Gani, a lake shore is dated at 21,000 bp (U.Q. 1149). Aterian unifacial pedunculate points generally line the paleobeaches.

The origin of the Aterian is given as the Maghreb for the northern Sahara, but Clark (1980) also indicates possible relationships of the Aterian in the southern Sahara with the Lupemban Industrial Complex known in the savannas of tropical Africa. Both processes imply migrations due to environmental change.

The dates proposed for the Aterian in northern Niger by Tillet (1983, 40,000–20,000 bp) and Durand *et al.* (1983, 26,000–20,000 bp), are consistent with the environmental data indicating favorable climatic change. The Aterian could be defined as the culture born and developed during Isotopic Stage 3.

The end of the Aterian is also related to climatic change. In the late Pleistocene a severe arid phase culminating at 18,000 bp correlates with the last glacial episode and the lowering of sea level and temperatures. Rainfall may have been reduced to as little as 20% of modern values, dune fields extended as far south as 10°N, and isohyets were located 500 to 700 km south of their modern position (Sarnthein 1978; Servant and Servant-Vildary 1980; Talbot 1980, 1984). Even the mountainous areas had sparse vegetation (Maley 1983). The central basins of the Sahara, today hyperarid, were then even more severely desertic. In the Taoudenni Basin this is represented by thick aeolian deposits. At 23°15'N, an Aterian site was found *in situ*, just beneath the lower layers of a 3 m thick paleodune which is capped by the earliest Holocene fossiliferous lake silts dated, at this latitude, to 8700 bp (Gif 6194) and 8600 bp (U.Q. 1142). This section demonstrates strikingly the causes of the end of the Aterian in the Sahara—it became impossible for plants or animals to survive and even today they are still nearly absent from those areas in which annual rainfall varies from 5 to 20 mm. The Aterians had to migrate. But to where? Mountains, river valleys, and lake shores are all classical refuges, but competition must have been severe even along the Niger, Senegal, and Nile rivers which suffered acute changes in their water balance (Michel 1973; Rognon 1976; Williams 1974). From the western and northern Sahara, man most probably migrated to the coasts of the Atlantic and Mediterranean where large areas of the continental plateau had emerged during the 120 m oceanic regression. Unfortunately, any of these sites which remain are now under sea level.

At *ca*. 20,000 bp in Morocco (Roche 1976) and Algeria (Camps *et al*. 1973), between 18,000 and 15,000 bp in Upper Egypt (Reed 1965; Lubell 1974), at 12,500 bp in Nubia (Anderson 1968), at 10,000 bp at Cape Juby (Petit-Maire 1973, 1979) and at 7000 bp in northern Mali (Dutour and Petit-Maire 1983; Petit-Maire and Dutour 1987; Dutour 1986), a new population appeared: the so-called "African Cromagnoids" or Mechta-Afalou group, possibly related to the Aterians. However, the available archaeological evidence does not support such a hypothesis: in the Maghreb, the Iberomaurusian follows the Aterian after a hiatus of as yet unknown duration and in the southern Sahara, the Ounanian appears after the aeolian gap noted previously. Roche (1963) denies any affinity between Aterian and Iberomaurusian, and Tillet (1983) sees no connection with the Ounanian. The physical anthropological data are somewhat more encouraging. Upper Pleistocene and Lower Holocene African Cromagnoids share several morphological traits with the few known Aterian remains (Ferembach 1976, 1985). They could have evolved quite rapidly, with environmental changes inducing modification in selective pressures, or increased cross breeding and drift as a consequence of increased migration. Cultural change could have paralleled genetic change. The combination of climatic deterioration and low sea level may have induced intercontinental migration or colonization of islands. Thoma (1978) and Ferembach (1985) boldly propose a common origin for African and European Cromagnoids, while Petit-Maire (1973, 1979) links the Izriten Mechtoids in southern Morocco with the Canary Guanches. It seems certain that some relationship exists between the alternating humid-arid phases in North Africa and variations in the biology of contemporaneous human populations.

The Holocene Interglacial and Neolithisation

Recent multidisciplinary studies have defined with great precision the Holocene environmental changes in the Taoudenni Basin of northern Mali, one of the most arid areas of the Sahara (Petit-Maire 1986, 1987; Petit-Maire and Riser 1981, 1983). In brief, a wide-spread lacustrine episode took place from 9000 to 4500 bp between 24° and 22°N, from 9500 to 4500 bp between 20° and 22°N, and from 9500 to 3500 bp between 18° and 20°N (Fig. 2.2). Lake or swamp deposits (Plate 2.II; Plate 2.III) are so extensive that the quantity of surface fresh water fed by local runoff (Fig. 2.3) implies precipitation varying from south to north from 600 to at least 300 mm in an area that today receives from 50 to 5 mm. After 7000 bp the climate progressively deteriorated and changes in lake balance became frequent and severe, leading to saline deposits or mud cracks (Plate 2.II) (Petit-Maire 1986). These data from an isolated hyperarid Saharan basin fit well with results obtained in numerous other areas from Ethiopia and Sudan to the Atlantic and from the Saoura to the Ghana lakes. They provide evidence that global change affected the paratropical hyperarid belts which probably disappeared completely during the Holocene.

This hypothesis is supported by new anthropological data from northern Mali where the remains of 117 individuals (Plate 2.IV) were collected from burials dated ca 7000 to 4500 bp. They have indisputably Cromagnoid features (Dutour and Petit-Maire 1983; Dutour 1986; Petit-Maire and Dutour 1987), thus suggesting a relationship with Mechta-Afalou populations of the Maghreb. We have already discussed their probable Aterian origin. From which refuges did they return during the Holocene? From the north, across a continuous steppe, or from the east or south, the Adrar or the Niger banks? Further research is needed to answer this question.

The lag shown in Fig. 2.3 between the dates for lake molluscs and those for paleosols, large herbivores and human skeletons, is striking. If the observed differences are not biased by the nature of the analysed materials, 1500 to 2000 years were necessary to provide a full biological response in a region which had been hyperarid for at least 10,000 years. Increased rainfall induced growth of a Gramineae steppe to the north of the area and of a Sudanese savanna south of the twentieth parallel. A complete trophic chain of foraminifera, ostracods, molluscs, large fish (up to 1.60 m), crocodiles (Plate 2.V), turtles, hippopotamus, rhinoceros, phacochoerus, elephants, and large antelopes (*Alcelaphus buselaphus, Limnotragus spekei, Hippotragus equinus*, etc.) was established (cf. Petit-Maire and Riser, eds. 1983:Plate 2.IV).

Man took advantage of the optimum environmental conditions to develop a new culture, the Neolithic (Petit-Maire 1985). The wide extension of a Gramineae steppe correlates with the ubiquity and density of mortars and grinders (Plate 2.IV). The abundance of large fish in shallow lakes must be related to the abundant production of bone hooks and harpoons (273 harpoons were found in a single site about 200 m² at 19°15'N; Decobert and Petit-Maire 1985). The possibility of a relatively sedentary life around clayey pans allowed the elaboration of a rich ceramics industry (Commelin 1984). The variety and quality of pottery decoration and of stone and bone adornments (Camps-Fabrer *et al.* 1982), suggest spare time for refinement, incompatible with a hard struggle for life in a hostile environment.

After 4500 bp, the archaeological sites are found on the dried lake beds, probably around the remaining water holes or wells at the center of the depressions. The large mammalian fauna is gone by 5000 bp and fresh water molluscs could no longer survive except in the Azawad (17°–19°N). Like the Aterians some 15,000 years earlier, man had to migrate. The populations of the eastern Taoudenni Basin moved to the Tilemsi valley (19°N) by 4000 bp where they mixed with negroid groups (Smith 1976, 1980; Petit-Maire and Dutour 1983; Dutour 1986). From inner Mauretania and the western Sahara, man migrated to the Atlantic coast, where fresh water holes and seafood allowed them to live until 2500 bp (Fig. 2.4). They also mixed with southern populations (Petit-Maire 1979, 1980).

At 3500 bp the Saharan basins were desert again and the interglacial humid phase came to an end, as had the preceeding one at around 100,000 bp.

The Present and the Future

The climatic evolution of the hyper-arid Saharan basins correlates well with the astronomical and oceanic isotopic curves (Fig. 2.5) for the last 150,000 years. Humid lacustrine or paludal phases also grossly correlate with the "warm" peaks of high atmospheric carbon dioxide (Lorius *et al.* 1985) while arid phases correlate with the "cold" ones. The origin and development of human cultures, promoted by favorable biotopes and sedentarization, are closely linked with these changes; and climatic landmarks clearly stake out the onset and the end of prehistoric civilizations. The Lower Paleolithic ended with the Mid-Pleistocene optimum (Isotopic Stage 5); the Aterian developed during the lacustrine episodes marking the end of Stage 3 and ended with the occurrence of aridity at Stage 2; the Neolithic began with the Holocene climatic optimum (Stage 1).

Such correlations between anthropology and paleoclimatology are still awkward due to the imprecision in dating lithic material. Such attempts should, however, be developed, since our future depends upon these relations: man is, unfortunately, one of the best climatic markers (Petit-Maire 1984) and paleoclimates have always been one of the fundamental factors in our evolution.

The astronomical curve (Fig. 2.5) calculated by Berger (1981) is the best prediction we have for long-term natural climatic evolution; the variations of terrestrial insolation will reach a new extreme about 60,000 years from now; a new glacial will probably set in and correlate once again with a major extension of the arid belts. The trend will be progressive and irregular in the future as it has been in the past and thus not be sensed at the human scale. The wide short-term oscillations (Fig. 2.6) of monsoonal and subtropical rains will still compensate one another until a dangerous threshold is reached when the dry years will no longer be balanced by the wet ones. This pattern of climatic evolution is being modified through the abrupt increase of anthropogenic carbon dioxide, which heats the atmosphere and thus creates a contrary danger (Lambert 1987). By all means, let us hope that the next climatic crisis—whatever it may be—will not correspond once more to a stagnation of culture.

References

Alimen, H., Beucher, F. and Conrad, G. 1966. Chronologie du dernier cycle Pluvial-Aride au Sahara nord-occidental. Paris: *C.R. Acad. Sci.* 263:5–8.

Anderson, J. 1968. Late palaeolithic skeletal remains from Nubia. In *Prehistory of Nubia* (ed F. Wendorf): pp. 996-1041. Dallas: SMU Press.

Aumassip, G. and Petit-Maire, N. 1982. Les industries préhistoriques de la vallée du Shati. In *Le Shati, Lac Pléistocène du Fezzan* (ed N. Petit-Maire): pp. 86–8. CNRS.

Berger, A. 1979. Insolation signature of Quaternary climatic change. *Il Nuovo Cimento* 26:63–87.

_____. 1981. Le soleil, le climat et leurs variations. *Ciel et Terre* 97:229-44.

_____. 1981. The astronomical theory of palaeoclimates. In *Climatic Variations and Variability: Facts and Theories:* pp. 501-25.

_____. 1984. Accuracy and stability of the Quaternary terrestrial insolation. *Milankovich and Climate* 1:83-111.

Berger, W. H., Killingley, J. S., Metzler, C. V and Vincent, E. 1985. Two step deglaciation. *Quaternary Research* 23:258-71.

Butzer, K. 1980. Pleistocene history of the Nile valley in Egypt and Lower Nubia. In *The Sahara and the Nile* (eds M. A. J. Williams and H. Faure): pp. 253-80. Rotterdam: Balkema.

Camps, G., Delibrias, G. and Thommeret, J. 1973. Chronologie des civilisations préhistoriques du Nord de l'Afrique d'après le radiocarbone. *Libyca* 21:65-89.

Camps-Fabrer, H., Commelin, D. and Petit-Maire, N. 1982. Objets rares et parures néolithiques du Sahara malien *Trav. LAPMO.* pp. 1-19.

Chaline, J. 1985. Histoire de l'homme et des climats au Quaternaire. Doin.

Chamard, P. 1973. Monographie d'une sebkha continentale du Sud-Ouest saharien: la sebkha de Chemchane (Adrar de Mauritanie). *Bull. IFAN* 35:207-43.

Chavaillon, N. 1971. L'Atérien de la Zaiouia-el-Kebira au Sahara Nord-Occidental. *Libyca* 19:9-51.

Chavaillon, N. and Fabre, J. 1960. L'Atérien d'El Guettara. *Bull. SPF* 5/6:346-54.

Clark, J. D. 1971. An archaeological survey of Northern Air and Ténéré. *Geogr. J.* 137:455-7.

Clark, J. D. 1980. Human populations and cultural adaptations in the Sahara and Nile during prehistoric times. In *The Sahara and the Nile* (eds M. A. J. Williams and H. Faure): pp. 527-82. Rotterdam: Balkema.

Commelin, D. 1984. *La céramique néolithique dans le bassin de Taoudenni (Sahara malien). Méthodes d'étude. Faciès géographiques et chronologiques. Relations avec les variations climatiques à l'Holocène.* (Thèse Doct. 3. Cycle, 1, Univ. Aix-Marseille II).

Conrad, G. 1969. *L'Évolution Continentale Post-Hercynienne du Sahara Algerien.* Paris: CNRS.

Decobert, M. and Petit-Maire, N. 1985. An early neolithic midden and necropolis in the Malian Sahara. *Nyame Akuma*: 26:26–32.

Duplessy, J. C. 1978. Isotope studies. In *Climatic Changes* (ed J. Gribbin): 46–67.

_____. 1984. Les calottes glaciaires. *La Recherche* 156:806–18.

Durand, A., Lang, J., Morel, A. and Roset, J. P. 1983. Évolution géomorphologique, stratigraphique et paléoclimatique au Pléistocène supérieur et à l'Holocène de l'Air oriental (Sahara méridional, Niger). *Rev. Géol. Dyn. et Geogr. Phys.*: 24:47–59.

Dutour, O. 1986. *Anthropologie écologique des populations néolithiques du Sahara malien.* (Thèse Aix-Marseille II, Luminy).

_____, and Petit-Maire, N. 1983. Sepultures et restes osseux. In *Sahara ou Sahel?* (eds N. Petit-Maire and J. Riser): pp. 275–92.

Fabre, J. 1983. Esquisse stratigraphique préliminaire des dépôts lacustres quaternaires de la région de Taoudenni-Thraza. In *Sahara ou Sahel?* (eds N. Petit-Maire and J. Riser): pp. 421–37.

_____, and Petit-Maire, N. 1983. Lacs pléistocènes de la région de Taoudenni (Sahara malien). In *Bassins Sedimentaires en Afrique*: pp. 101–2. Marseille-St. Jerome: Trav. Lab. Sc. Terre.

Faure, H. 1962. Reconnaissance géologique des formations sédimentaires post-paléozoïques du Niger oriental. *BRGM*: pp. 36-46.

_____, and Gac, J. Y. 1981. Will the Sahelian drought end in 1985? *Nature* 291:475–8.

Ferembach, D. 1976. Les restes humains atériens de Temara. *Bull. Mem. Soc. Anthr.* 3:175–80. Paris.

_____. 1976. Les restes humains de la grotte de Dar-es Soltane II. *Bull. Mem. Soc. Anthr.* 3:183–93. Paris.

_____. 1985. On the origin of the Iberomaurusians. A new hypothesis. *J. Human Evol.* 14:393–7.

Gasse, F. 1975. *L'évolution des lacs de l'Afar central du Plio-Pléistocène à l'Holocène.* (Thèse, Paris).

———, Rognon, P. and Street, A. 1980. Quaternary history of the Afar and Ethiopian Rift lakes. In *The Sahara and the Nile* (eds M. A. J. Williams and H. Faure): pp. 361-400. Rotterdam: Balkema.

Gaven, C. 1982. Radiochronologie isotopique uranium-thorium. In *Le Shati, Lac Pléistocène du Fezzan* (ed N. Petit-Maire): pp. 44–54. Editions CNRS.

———, Hillaire-Marcel, C. and Petit-Maire, N. 1980. A Pleistocene lacustrine episode in southeastern Libya. *Nature* 290:131–3.

Hays, J. D., Imbrie, J. and Shackleton, N. J. 1976. Variations in the earth's orbit: pacemaker of the ice ages. *Science* 194:1121–32.

Jakel, D. 1977. Abflurs und Formungsvorgänge in Tibesti-Gebirge. *X° Congrès INQUA*. Birmingham.

Lambert, G. 1987. Le gaz carbonique dans l'atmosphère. *La Récherche* 189:778–87.

Livingstone, D. 1980. Environmental changes in the Nile headwaters. In *The Sahara and the Nile* (eds M. A. J. Williams and H. Faure): pp. 339-60. Rotterdam: Balkema.

Lorius, C., Jouzel, J., Ritz, C., Merlivat, N., Barkov, N., Korotkevich, Y. S. and Kotlyakov, V. M. 1985. A 150,000 years climatic record from Antarctic ice. *Nature* 316:591–6.

Lubell, D. 1974. The Fakhurian. A late palaeolithic industry from northern Egypt. *Geol. Survey Egypt* 58.

McBurney, C. B. 1967. *The Haua Fteah (Cyrenaica) and the Stone Age of the South-East Mediterranean.* Cambridge: Cambridge University Press.

Maley, J. 1981. Etudes palynologiques dans le Bassin du Tchad et paléoclimatologie de l'Afrique nord-tropicale de 30,000 ans à l'époque actuelle. *Doc. ORSTOM* : Paris.

———. 1983. Histoire de la vegetation et du climat de l'Afrique nord-tropicale au Quaternaire récent. *Bothalia* 14:377–389.

———, Roset, J. P. and Servant, M. 1971. Nouveaux gisements préhistoriques au Niger oriental. *Bull. ASEQUA* 31:9–18.

Michel, P. 1973. Les bassins des fleuves Sénégal et Gambie, étude géomorphologique. *Mem. ORSTOM*: Paris.

Petit-Maire, N. 1973. Occupation humaine holocène de la région du Cap Juby (Sud-Ouest marocain). *Bull. Mem. Soc. Anthr.* 13 (10):379–412. Paris.

———. 1979. Les formations du Wadi Shati. *Bull. AFEQUA*: 56–57:77–78.

———. (ed) 1979. *Le Sahara Alantique à l'Holocène. Peuplement et Écologie.* CRAPE, Alger, Mem. XXVIII.

———. (ed) 1982. *Le Shati, Lac Pléistocène du Fezzan.* CNRS. Marseille.

———. 1984a. L'Homme, marqueur paleoclimatique. *Géochroniques* 11:13–15.

———. 1984b. Le Sahara, de la steppe au désert. *La Recherche* 160:1372–1382.

———. 1984c. La néolithisation au Sahara: problèmes chronologiques, géographiques et paléoclimatiques. *Cah. ORSTOM* 14(2): 189–212.

———. 1985. Dynamique sédimentaire récente du Bassin de Taoudenni. *Abstracts Geol. Soc. Afr.* Gaborone.

———. 1986a. *Homo climaticus*: vers une paléoanthropologie écologique. *Bull. Soc. Roy. Belge Anthr. et Préhist.* 97:59–75.

———. 1986b. Palaeoclimates in the Sahara of Mali. A multidisciplinary study. *Episodes* 9-1:7–15.

———, Celles, J. C., Commelin, D., Delibrias, G. and Raimbault, M. 1983. The Sahara in northern Mali: Man and his environment between 10,000 and 3500 years bp. *Afr. Arch Rev.* 1:105–125.

———, Delibrias, G., and Gaven, C. 1980. Pleistocene lakes in the Shati area, Fezzan. *Palaeoecology of Africa* 12:289–295.

———, and Dutour, O. 1987. Holocene populations of the western and southern Sahara: Mechtoids and paleoclimates. In *Prehistory of Arid North Africa* (ed A. Close): pp. 259–86. Dallas: SMU Press.

———, and Riser, J. 1981. Holocene lake deposits and palaeoenvironments in northeastern Mali. *Palaeogeogr., Palaeoclimatol., Palaeoecol.* 35:45–61.

———, and Riser, J. (eds) 1983. *Sahara ou Sahel? Quaternaire recent du Bassin de Taoudenni.* Marseille.

Reed, C. 1965. A human frontal bone from the late Pleistocene of the Kom Ombo Plain, Upper Egypt. *Man* 65:101–104.

Roche, J. 1963. *L'Épipaléolithique Marocain*. 2 vol. Paris: Didier.

_____. 1976. Cadre chronologique de l'épipaléolithique marocain. *IX° Congrès Sc. Préh*. Nice.

Rognon, P. 1976. Essai d'interprétation des variations climatiques au Sahara depuis 40,000 ans. *Rev. Geol. Dyn. et Geogr. Phys.* 18:251–282.

_____. 1980. Une extension des déserts (Sahara et Moyen Orient) au cours du Tardiglaciaire. *Rev. Geol. Dyn. et Geogr. Phys.* 22:313–328.

Sarnthein, M. 1978. Sand deserts during glacial maximum and climatic optimum. *Nature* 272:43–46.

Servant, M. 1973. *Séquences continentales et variations climatiques: évolution du Bassin du Tchad au Cénozoique supérieur*. (These, Paris).

_____, and Servant-Vildary, S. 1980. L'environnement quaternaire du Bassin du Tchad. In *The Sahara and the Nile* (eds M. A. J. Williams and H. Faure): pp. 133-62. Rotterdam: Balkema.

Shackleton, N. 1981. Paleoclimatology before our ice age. In *Climatic Variations and Variability* (ed A. Berger): pp.167–80.

_____. 1982. The deep sea sediment records of climate variability. *Prog. Oceanogr.* 2:199–218.

Smith, A. 1979. Biogeographical considerations of colonization of the lower Tilemsi valley in the second millenium B.C. *J. Arid Environments* 2:355–361.

Smith, A. B. 1980. The neolithic traditions in the Sahara. In *The Sahara and the Nile* (eds M. A. J. Williams and H. Faure): pp. 451-66. Rotterdam: Balkema.

Talbott M. 1980. Environmental responses to climatic change in the West African Sahel over the past 20,000 years. In *The Sahara and the Nile* (eds M. A. J. Williams and H. Faure): pp. 37-62. Rotterdam: Balkema.

_____. 1984. Late Pleistocene rainfall and dune building in the Sahel. *Palaeoecol. of Africa* 16:203–214.

Thoma, A. 1978. L'origine des Cromagnoïdes. In *Les Origines Humaines et les Époques de l'Intelligence* (ed J. Piveteau): pp. 261–282. Paris: Masson.

Tillet, T. 1983. *Le Paléolithique du Bassin Tchadien Septentrional*. Marseille: CNRS.

Williams, M. and Adamson, D. 1974. Late Pleistocene dessication along the White Nile. *Nature* 248:584–586.

Williams, M. A. J. and Adamson, D. A. 1980. Late Quaternary depositional history of the Blue and White Nile rivers in Central Sudan. In *The Sahara and the Nile* (eds M. A. J. Williams and H. Faure): pp. 281-304. Rotterdam: Balkema.

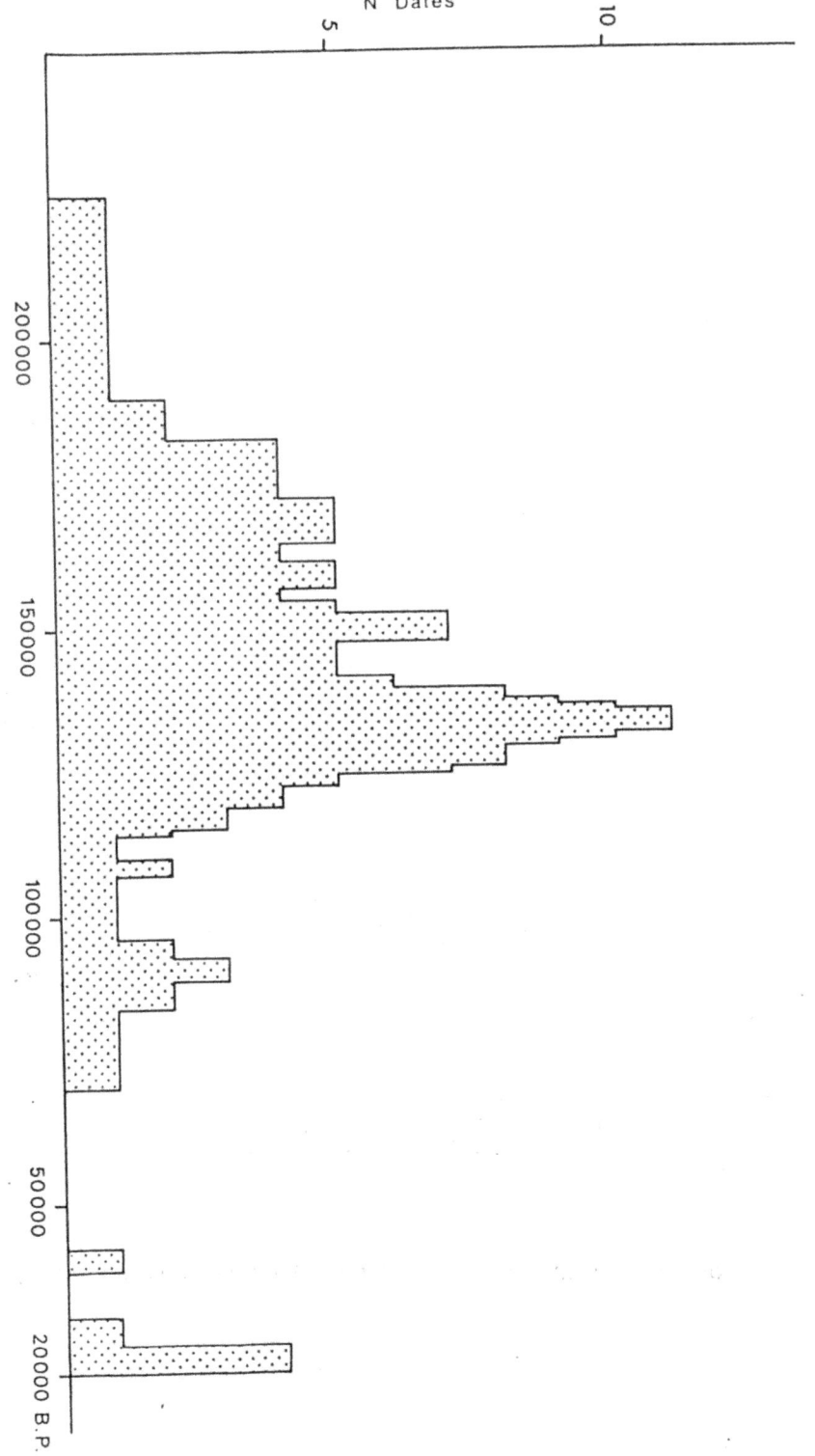

Fig. 2.1: Th/U and 14C ages for the Shati valley lake.

Climatic Change and Man in the Sahara

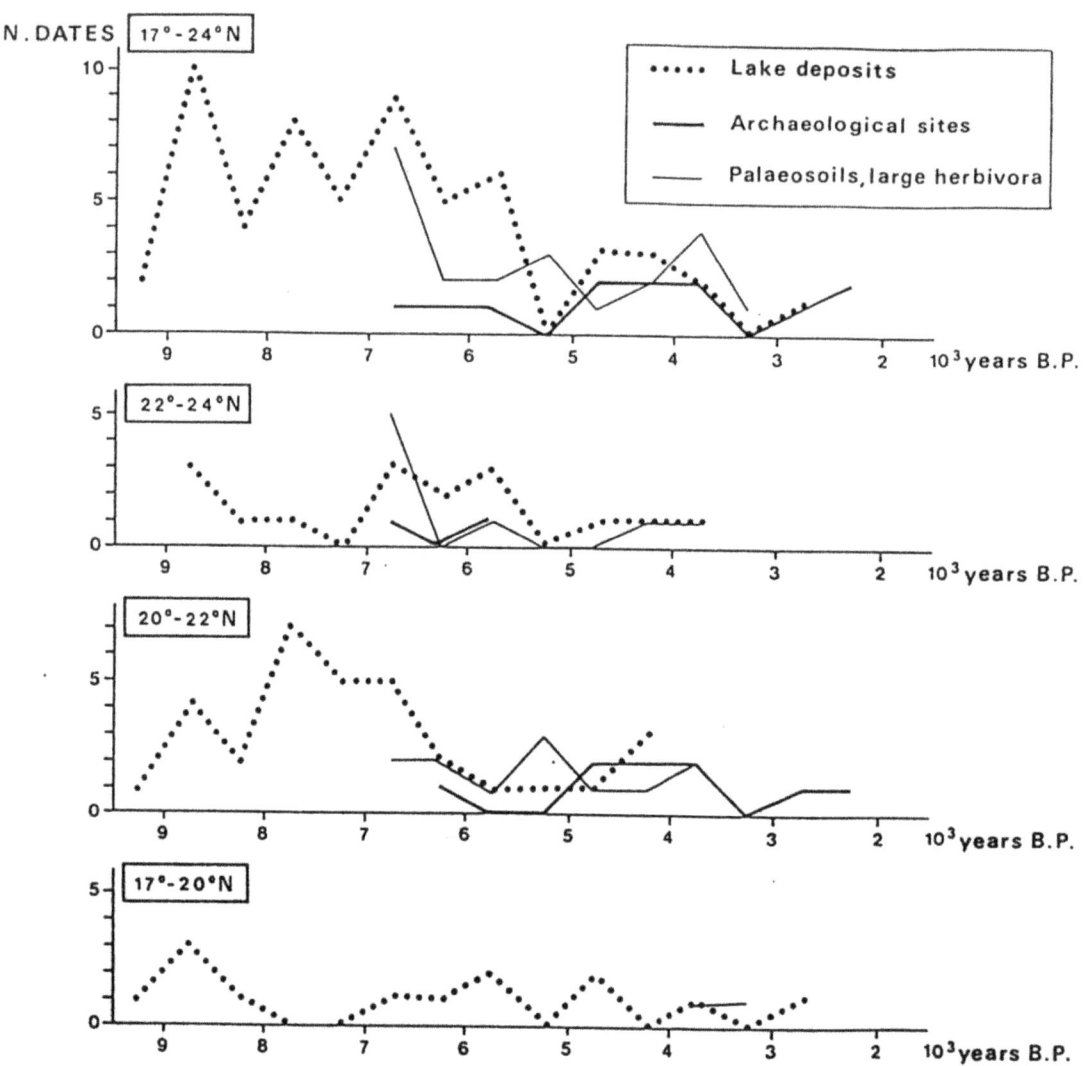

Fig. 2.2: Radiocarbon ages (Gif and U.Q.) for the Holocene of malian Sahara.

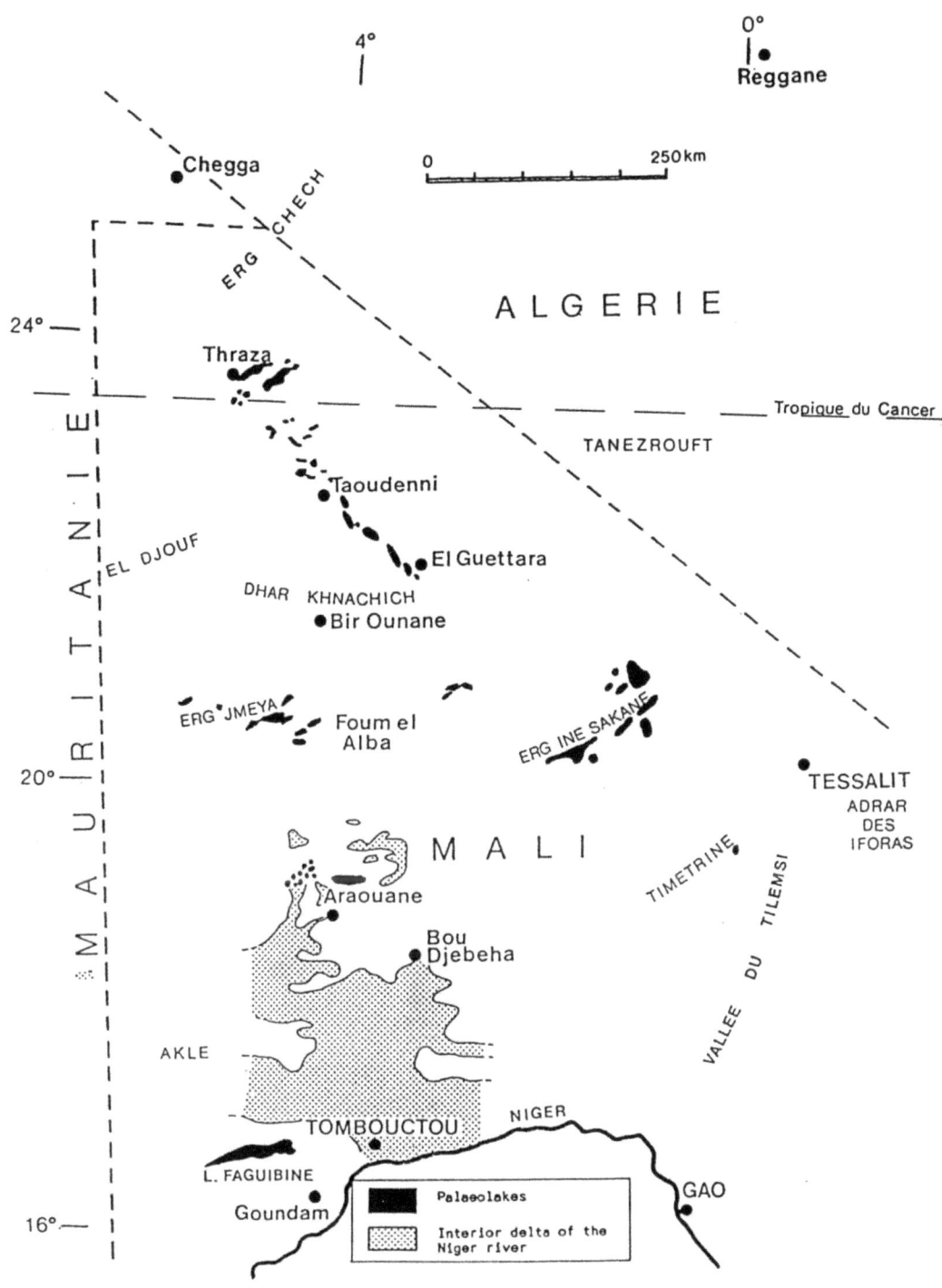

Fig. 2.3: Holocene lakes and northern interior delta of the Niger in northern Mali.

Climatic Change and Man in the Sahara

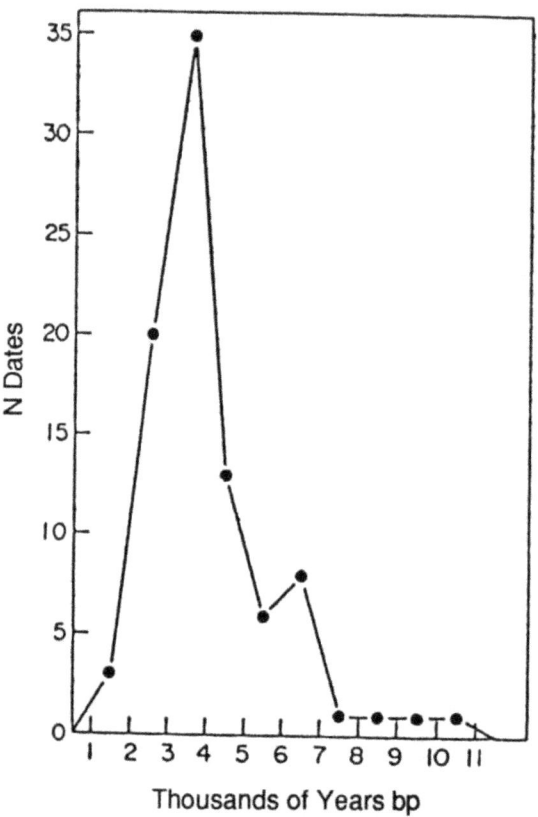

Fig. 2.4:   Radiocarbon ages for Holocene human populations along the Atlantic coast of the Sahara.

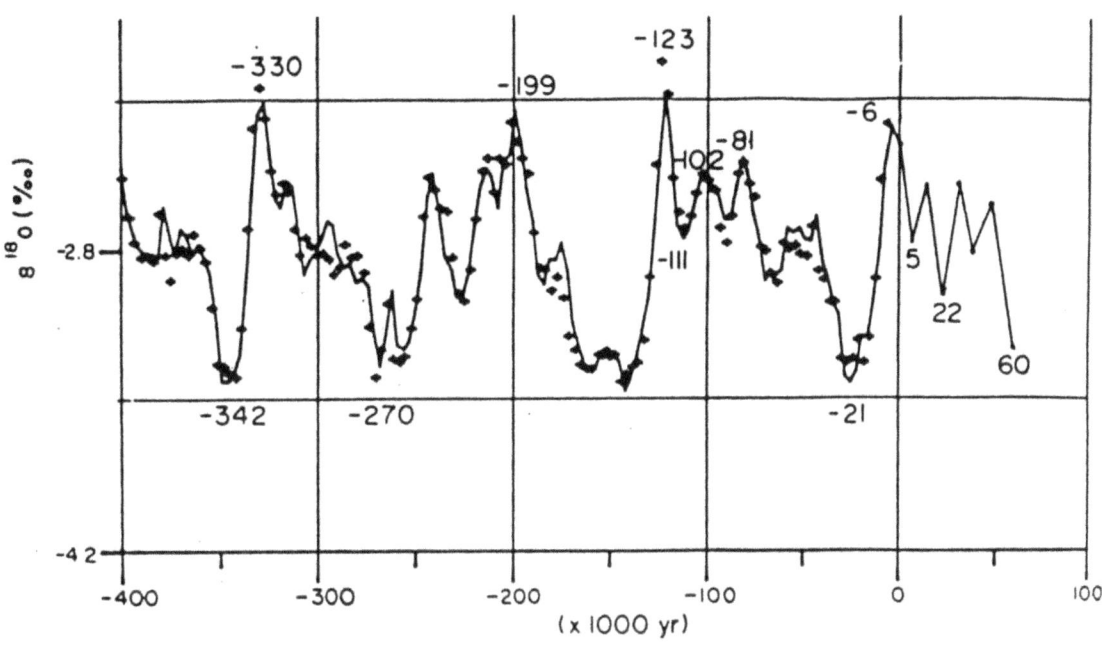

Fig. 2.5: Long-term climatic variations over the past 400,000 years and predictions for the future (after A. Berger 1981:518).

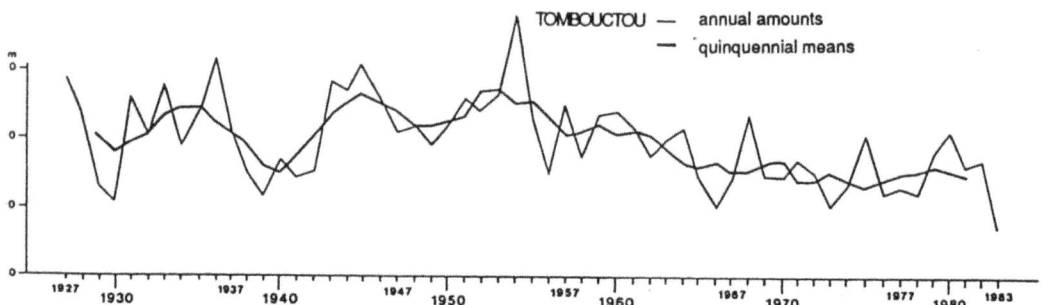

Fig. 2.6: Annual rainfall and mobile quinquennial means at Tombouctou since 1927.

Plate 2.I: The Lake Shati mollusc thanatocoenosis.

Climatic Change and Man in the Sahara

Plate 2.IIa: Holocene lake deposits at Wadi Haijad.

Plate 2.IIb: Mud-cracks level in lake deposits (seen in plane).

Plate 2.IIIa: Section showing Late Pleistocene eolian layers under Holocene lake deposits, Sbeita.

Plate 2.IIIb: Lakelet deposits in interdune trough, Umm el Assel.

Climatic Change and Man in the Sahara

Plate 2.IIIc: Section with successive swamp, lake and crust layers, Sbeita.

Plate 2.IIId: Swamp deposits, Telig.

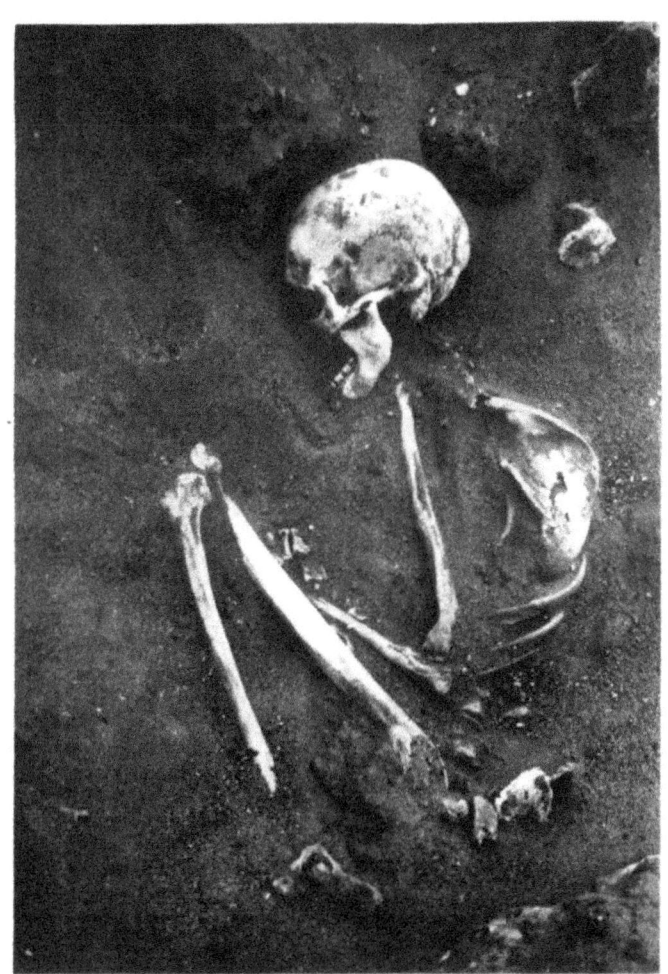

Plate 2.IVa: Burial, Erg Ine Sakane.

Plate 2.IVb: Grinding artifacts, Erg Jmeya.

Climatic Change and Man in the Sahara

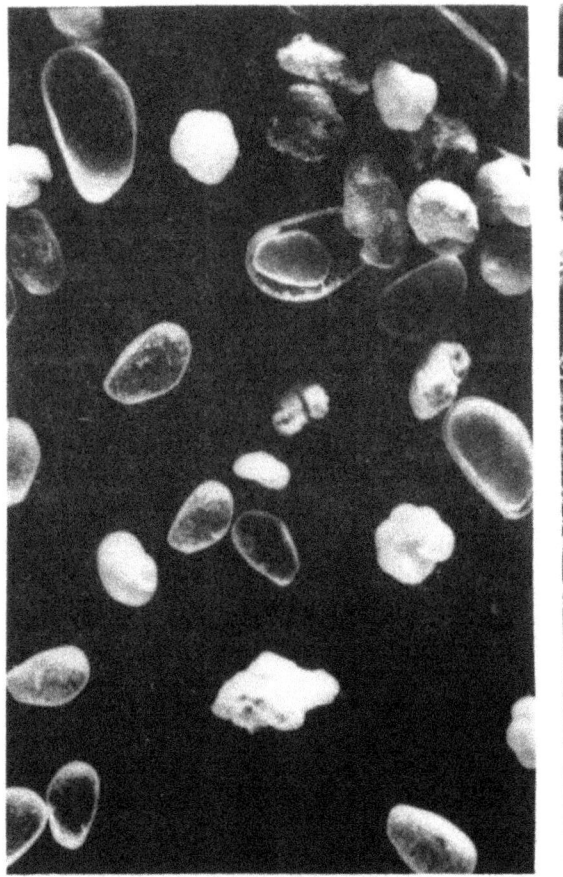

Plate 2.Va: Foraminifera and ostracods in lake silts.
Plate 2.Vb: *Melania tuberculata*, *Bulinus truncatus*, and *Biomphalaria pfeifferi* in lake deposits.

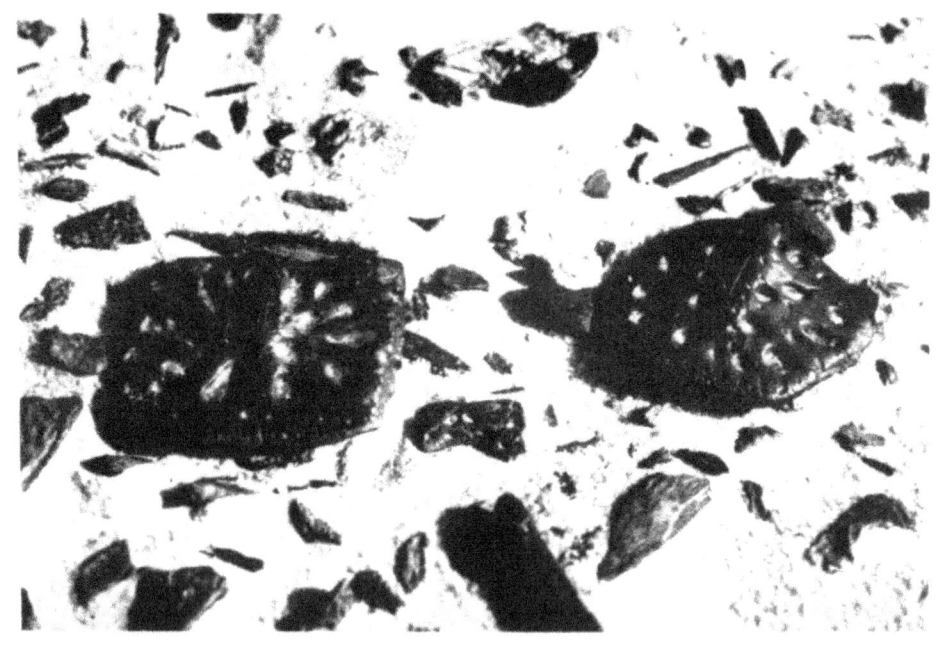

Plate 2.Vc: *Crocodylus niloticus* dermic plates, Hassi-el-Abiod.

After the Deluge: The Neolithic Landscape in North Africa

By Martin A. J. Williams, *Department of Geography and Cenozoic Research Unit, Monash University, Clayton, Victoria, Australia.*

Introduction

In spite of a geographically uneven distribution of carefully investigated sites (Fig. 3.1). leading to inevitable spatial and temporal gaps, the record of late Quaternary environmental fluctuations in North Africa is still the most complete and best dated of anywhere in the tropics. Critical evaluations of the physical and biological evidence used in the reconstruction of late Quaternary prehistoric environments in North Africa include recent comprehensive reviews by Adamson (1982), Street-Perrott *et al.* (1985), Williams (1984, 1985a) and Williams *et al.* (1986). The evidence itself ranges from geomorphic studies of rivers, dunes, lakes, and mountains through paleoecological investigations of micro- and macro-fossils of plants and animals (pollen, diatoms, molluscs, bones) to geochemical and isotopic analyses of organic and inorganic carbonates, paleosols, and groundwater. Although late Quaternary climatic changes were broadly synchronous in North Africa, the physical and biological impact of such changes was often time-transgressive, as was the cultural response, evident for instance in the first recorded appearance of domesticated plants and animals at Neolithic sites in North Africa.

From Late Pleistocene Aridity to Early Holocene High Lake Levels

The often well-vegetated fixed dunes that extend for over 5000 km from the mouth of the Senegal River in the west to the alluvial plains of the Sudanese Gezira in the east were mobile and probably devoid of vegetation during the very late Pleistocene, when the southern margin of the Sahara extended up to 500 km closer to the equator than it does today (Grove 1958, 1985; Sarnthein 1978; Mainguet *et al.* 1980; Talbot 1980). At this time, which coincided with the peak of the last glaciation and its immediate aftermath, considerable quantities of desert dust were blown westward from the Sahara and Chad basin to accumulate far out to sea in the ocean depths of the equatorial Atlantic (Parkin and Shackleton 1973; Kolla *et al.* 1979; Sarnthein *et al.* 1981). Separate dust plumes travelled from the northern Sahara toward the Mediterranean and Atlantic, depositing en route a mixture of quartz and calcareous silts in sheltered valleys such as those of the Tunisian Matmata massif (Coudé-Gaussen and Rognon 1983; Coudé-Gaussen 1984; Paquet *et al.* 1984). The obvious inference—that winds were stronger and the climate drier than today—is supported by the evidence of very low lake levels in central and eastern North Africa between about 18,000 and 11,000 years bp (Street-Perrott *et al.* 1985:Fig. 8.5).

After the Deluge: The Neolithic Landscape in North Africa

Fig. 3.2 and 3.3 are compilations of late Quaternary hydrological events from a number of reasonably well-dated localities in North Africa, including Ethiopia and the Nile valley. As these figures clearly illustrate, there is a striking contrast between the very high lake levels of the early Holocene and the long interval at the close of the Pleistocene when many of the North African lakes were low or dry. In the case of Lake Chad (which receives much of its water from the equatorial Adamawa Mountains of Cameroon), Lake Abhe (a terminal basin of the Awash River, located in the presently arid Afar Rift but fed by runoff from the Ethiopian uplands), and Lakes Ziway and Shala in the now semi-arid Ethiopian Rift, this late Pleistocene arid interval probably lasted for over 5000 (radiocarbon) years and was succeeded by two or more early to mid Holocene wet intervals, each about 2000 years in duration.

In the high mountains of East Africa regional snow lines were 800–1200 m lower at the height of the last glaciation than they are today (Livingstone 1980; Messerli and Winiger 1980). Periglacial solifluction was active well below present limits, and the upper timberline was correspondingly about 1000 m lower. Following deglaciation temperatures rose rapidly and by the start of the Holocene were at least as warm as today except perhaps in the Chad Basin, where some of the diatom assemblages could indicate cooler temperatures (Servant and Servant-Vildary 1980).

Toward the end of the Pleistocene, lake levels began to rise rapidly throughout the northeast quadrant of Africa, including the equatorial headwater region of the White Nile and the upland Ethiopian catchments of the Blue Nile, Atbara and Sobat rivers (Butzer *et al.* 1972; Gasse 1975; Street and Grove 1976; Perrott 1979; Williams and Adamson 1980). The level of Lake Victoria rose until it overflowed into the White Nile at about 12,500 years bp (Kendall 1969), no doubt flooding an extensive area of what are now the Sudd swamps of the southern Sudan. Flow in the Blue Nile was augmented by greatly increased runoff from the Ethiopian highlands towards 12,000–11,500 bp (Williams and Adamson 1980). Lake Turkana overflowed across an ill-defined divide to bring more water to the White Nile via the Lotigipi Swamp and the Sobat River (Butzer 1980a; Adamson and F. M. Williams 1980:232–5; Harvey and Grove 1982). The outcome of such exceptionally high rates of flow in the Blue and White Nile rivers was widespread flooding and clay deposition along their distal flood plain reaches in the Gezira plains of the central Sudan (Tothill 1946; Williams 1966), and the unprecedented "wild Nile floods" in Egypt toward 11,500–11,000 years bp (Butzer 1980b). This order of magnitude increase in the discharge of the Nile triggered and was accompanied by the accumulation of dark sapropelitic organic muds in the eastern Mediterranean (Stanley and Maldonado 1977; Adamson *et al.* 1980; Rossignol-Strick *et al.* 1982).

Situated some 2000 km west of the confluence of the Blue and White Nile rivers, Lake Chad was also high towards 11,000 years bp, as was Lake Bosumtwi, a crater lake in the present-day rain forest zone of Ghana (Servant and Servant-Vildary 1980; Talbot and Delibrias 1980). Both lakes recorded a short, intense regression around 10,500 bp (as did Lake Victoria) but rose again to very high levels during 9000–8000 bp. Between Lake Chad and the Nile, a number of small lakes in the Sudanese provinces of Darfur and Kordofan were high between >8500 bp and *ca.* 6000 bp (Fig. 3.3). Only a single major early Holocene high lake phase is evident at Oyo, En Nahud, and Wad Mansurab, after which these lakes dried out for good. During this same time other sites near Oyo show two distinct lake phases (Gabriel and Kröpelin 1983, 1984), demonstrating that some lakes are characterised by a more sensitive response to fluctuations in groundwater level, runoff, and evaporation than others within the same general region. It is therefore unwise to generalise unduly about regional fluctuations in climate from one or two isolated lake sites, particularly when it is not known whether at any one time such lakes are in fact amplifying or masking the regional climatic oscillations.

Mid to Late Holocene Desiccation

The regression which immediately followed the 9000–7000 bp interval of high North African lake levels is of considerable interest to prehistorians because at a number of places it seems to herald the closing stages of the widespread and remarkably uniform Epi-Paleolithic hunter-gatherer-fisher or "aqualithic" tradition (Sutton 1977) of the southern Sahara and Nile valley, including the N'Kiffi hippo-hunting culture of Adrar Bous (Clark, Williams, and Smith 1973) and the Early Khartoum tradition (Arkell 1949), which is now securely dated to 8000–7000 bp (Adamson, Clark, and Williams 1974).

Although lake levels rose again at a number of North African sites during the latter half of the Holocene (Fig. 3.2 and 3.3), the pastoral Neolithic inhabitants of the Sahara were dealing with an environment very different from that which gave rise to the ubiquitous barbed bone points used by their predecessors to spear the aquatic faunas of the lake-studded Sahara. No longer *le Sahara des lacs*, it was now the Sahara (and Nile) of *après le déluge*. Put succinctly, until *ca.* 7000 bp the Holocene climate of North Africa was wetter than today, albeit interspersed with drier intervals; thereafter it was mainly dry, with short-lived, moister intervals.

After the Deluge: The Neolithic Landscape in North Africa

Fig. 3.4 and 3.5 show the present-day rainfall and vegetation zones in North Africa. Leaving aside those mountainous areas where the climate and plant cover are moderated by altitude, the overall pattern is surprisingly simple, with precipitation decreasing rapidly away from the equator and away from the Mediterranean, and vegetation types arranged in broad latitudinal zones. The southern limit of active Saharan dunes more or less coincides with the 100 mm isohyet (Fig. 3.4). The fixed Pleistocene dunes of the sahel zone occupy what is today savanna thornscrub and savanna woodland with a rainfall generally in excess of 150 mm a year but usually less than 400–500 mm (Fig. 3.5). A simple comparison of the southern limit of presently fixed and active dunes (Fig. 3.4 and 3.5) suggests that during the peak of late Pleistocene aridity the southern limit of the Sahara lay up to 500 km closer to the equator than it does today. During the early Holocene the converse was probably true, with a broad zone of semi-desert scrub and grassland extending to the footslopes of the central Saharan uplands, allowing the large tropical herbivores to graze at least seasonally in what is now the heart of the Sahara. It is also likely that the broad littoral zone of Mediterranean steppe and forest also extended well south into the desert, probably as fingers of riparian woodland along major rivers like the Saoura which carried floodwaters from the Atlas Mountains well over a thousand kilometres south into the desert. Early Holocene pollen spectra from the south-central Sahara are characterised by both Mediterranean and tropical floral elements (Quézel and Martinez 1958, 1962), suggesting that both winter and summer rains extended well beyond their present limits.

The predictable impact upon the Neolithic landscape of North Africa of the mid to late Holocene hydrological changes illustrated in Fig. 3.2 and 3.3 was a retreat of the Mediterranean and savanna floras of the central Sahara back to their respective heartlands (Fig. 3.5). The monsoonal summer wet season became shorter and the rains more erratic. Groundwater levels fell as runoff from the uplands decreased, thereby depriving the myriad lakes of Niger, Chad, and the western Sudan of their prime source of supply. Hitherto perennial rivers became seasonal, seasonal streams became ephemeral, and reactivated dunes migrated across former river valleys and now dry lake floors.

In the Gezira plains of the central Sudan, progressive downcutting by the Blue Nile diverted water from the distributary channels that once flowed across these alluvial plains, so that the seasonally flooded toich lands began to dry out, and land snails like *Limicolaria flammata* began to colonise the now inactive Gezira flood plain toward 5000–4000 bp. As with any complex geomorphic system, the response of rivers, lakes, and dunes to the initial climatic perturbations which triggered these changes was variable and often time-transgressive (cf. Vita-Finzi

M. A. J. Williams

1973:82). For example, the final drying out of the extensive swamps fringing the lower White Nile was not achieved until long after the White Nile had begun to cut down into its flood plain in response to early to mid Holocene entrenchment of the Blue Nile (Adamson, Gillespie and Williams 1982). Swampy embayments persisted until at least 2700 bp near Shabona prehistoric site, over a hundred kilometres upstream of Khartoum, and until *ca*.1900–1500 bp near Jebel et Tomat prehistoric site, over twice that distance upstream of Khartoum (Fig. 3.5).

What might have been the cultural response of Neolithic communities to these climatically triggered but diachronous changes in the North African landscape? And, equally interesting, what influence did these Neolithic farmers and pastoralists in turn exert upon their biophysical environment?

Diachronous Inception of Neolithic Food Production in North Africa

Fig. 3.6 shows the first recorded appearance of certain domesticated plants and animals at a number of reasonably well-dated sites in North Africa. From west to east the sources of data are as follows: Tichitt (Munson 1976); Acacus (Mori 1965); Adrar Bous (Clark, Williams, and Smith 1973); Nabta (Wendorf and Schild 1980); Nile valley (Clark 1980; Hassan 1980); Khartoum (Krzyzaniak 1980); Tomat (Clark and Stemler 1975); and Besaka (Brandt 1980; Clark 1980). Four main conclusions may be drawn from the data illustrated in Fig. 3.6.

First, the inception of herding and of cereal domestication was strongly time-transgressive. Cattle were being herded during the early Holocene at Nabta playa and nearby Bir Kiseiba (Wendorf, Schild, and Close 1984) and in the Acacus massif, during the mid Holocene at Adrar Bous and at Khartoum, but not until the late Holocene at Dhar Tichitt, Jebel et Tuweimat (=Tomat) and Lake Besaka. Cereal crops were no exception. Domesticated barley first becomes evident *ca.* 6500 bp at farming sites in the Nile valley—or over a thousand years *after* barley was being cultivated at Nabta playa and other sites in the western desert of Egypt. In contrast to the thinner-stemmed temperate festucoid cereal grasses like wheat and barley, the tough thick-stemmed tropical panicoid grasses like millet and sorghum were not cultivated until comparatively late in the Holocene. It may be, as Stemler (1980) suggests, that late development of an efficient harvesting tool might have delayed domestication of millet and sorghum. Equally, swampy clay soils may have impeded sorghum cultivation at Jebel et Tomat until about 1900 bp, and ready availability of wild cereal grasses may have minimised the need to grow millet at Dhar Tichitt until *ca.* 3000 years bp.

Second, initial domestication of plants and animals was not necessarily synchronous at any one locality. At Dhar Tichitt cattle were being herded some 500 years before millet was first cultivated, and at Nabta playa and Bir Kiseiba cattle appear in the archaeological record nearly 1500 years before domesticated barley (Wendorf, Schild and Close 1984).

Third, and as a rider to the first two conclusions, it is evident that the onset of early Neolithic food production in North Africa cannot be correlated with any one type of climatic event. At certain sites the emergence of herding or cultivation coincides with a change from wetter to drier climate, at others from drier to wetter. The Neolithic pastoralists at Adrar Bous could water their cattle at a lake that was comparatively fresh and deep. The millet cultivators at Dhar Tichitt, on the other hand, had to rely upon local rainfall to water their crops, for by then the Tichitt lake had finally dried out.

Finally, it is possibly significant that if time lines were drawn linking these sites, they would resemble a bow wave advancing from the north and fanning out to west, east and south at about the time that increasing desiccation was forcing the Neolithic inhabitants of the Sahara to migrate into regions that until then had been too wooded and too tsetse fly–ridden for the comfort and well-being of both herders and their cattle.

Neolithic Impact upon the North African Landscape

Severe droughts have afflicted North Africa on four occasions this century: 1913, the early 1940s, 1968–1974, and the 1980s. One outcome of the more recent droughts was an increase in the number of dust storms—an increase which has provoked speculation that humanly induced degradation of the Sahelian vegetation ("desertification") brings about a further reduction in rainfall via the increase in surface reflectivity or albedo (Charney 1975). The argument is unconvincing, for it fails to explain why the droughts end, and it takes no account of the fact that during the past hundred years, at least, major droughts in Ethiopia, India, and Australia have all been synchronous, indicating worldwide rather than purely local climatic oscillations. One can therefore reject the false but appealing notion that the aridity that provoked the Neolithic exodus from the Sahara from about 4500–4000 bp onward was itself engendered by Neolithic overgrazing. Furthermore, Australia has also become drier since the mid Holocene, but cattle did not reach inland Australia until the last century.

At one or two sites, such as Adrar Bous, there seems to be an order of magnitude increase in sedimentation rates during the time of Neolithic settlement relative to the early Holocene. Here again, it is not possible to implicate humans and their herds rather than a decrease in plant cover brought about by an overall reduction in precipitation after the moist interval recorded in high lake levels towards 9000–8000 bp.

Despite the preceding caveats there does appear to be evidence that Neolithic herding and farming have indeed brought about accelerated and locally irreversible erosion and land degradation in North Africa. Along the middle Awash valley of Ethiopia's arid Afar Rift there is a thick sequence of horizontal fluvio-lacustrine Plio-Pleistocene sediments. Gully erosion, active today, has reduced much of the area to a typically "badlands" topography. The *kerrib* or Sudanese badlands bordering the Atbara and Blue Nile rivers are active today. Much further afield, in Tunisia, the horizontally stratified loess mantles that accumulated in the sheltered valleys of the Matmata massif are today severely dissected by gully erosion. In north-central India, the Son and Belan valleys are lined with horizontally stratified alluvial clays and sands and loess sheets of late Pleistocene age. These deposits are also severely eroded by active gullies or *nalas*. In every instance, the gully erosion is a Holocene phenomenon, for cut-and-fill structures of a similar size and extent to the badland gullies are totally absent from the well-exposed Pleistocene deposits. We can reject two possible causes—tectonic activity and climatic change—since neither can explain the lack of badland erosion in these regions during the Pleistocene. By default, anthropogenic erosion seems the only satisfactory explanation.

A clue to what might have happened comes from Australia. Many of the alluvial valley fills of southeastern Australia are now badly gullied, and many of the gullies are less than two centuries old. During the heavy rains that often herald the termination of a drought, cattle and sheep tracks concentrate runoff, rills form, and some of these develop within a few short years into deep, narrow, active gully systems. In view of this experience, it seems more than probable that Neolithic herding of hoofed animals, and clearing of grass and trees to grow crops, have been responsible for initiating the wave of accelerated erosion that seems to have become a feature of the North African landscape from mid Holocene times onwards.

References

Adamson, D.A. 1980. The integrated Nile. In *The Sahara and the Nile* (eds M. A. J. Williams and H. Faure): pp. 221–34. Rotterdam: Balkema.

Adamson, D., Clark, J. D. and Williams, M. A. J. 1974. Barbed bone points from central Sudan and the age of the "Early Khartoum" tradition. *Nature* 249:120–3.

Adamson, D. A., Gasse, F., Street, F. A. and Williams, M. A. J. 1980. Late Quaternary history of the Nile. *Nature* 288:50–5.

Adamson, D. A., Gillespie, R. and Williams, M. A. J. 1982. Palaeogeography of the Gezira and of the lower Blue and White Nile valleys. In *A Land between Two Niles: Quaternary Geology and Biology of the Central Sudan* (eds M. A. J. Williams and D. A. Adamson): pp. 165–219. Rotterdam: Balkema.

Arkell, A.J. 1949. *Early Khartoum*. London: Oxford University Press.

Atlas of Africa. 1973. Paris: Editions Jeune Afrique.

Brandt, S. A. 1980. Archaeological investigations at Lake Besaka, Ethiopia. In *Proceedings of the 8th Panafrican Congress on Prehistory and Quaternary Studies* (eds R. E. Leakey and B. A. Ogot): pp. 239–43. Nairobi: TILLMIAP.

Butzer, K. W. 1980a. The Holocene lake plain of North Rudolf, East Africa. *Physical Geography* 1:42–58.

Butzer, K. W. 1980b. Pleistocene history of the Nile valley in Egypt and lower Nubia. In *The Sahara and the Nile* (eds M. A. J. Williams and H. Faure): pp. 253–80. Rotterdam: Balkema. .

Butzer, K. W., Isaac, E. L., Richardson, J. L. and Washbourn-Kamau, C. 1972. Radiocarbon dating of East African lake levels. *Science* 175:1069–1076.

Charney, J.G. 1975. Dynamics of deserts and drought in the Sahel. *Quart. J. Roy. Met. Soc.* 101:193–202.

Clark, J.D. 1980. The origins of domestication in Ethiopia. In *Proceedings of the 8th Panafrican Congress on Prehistory and Quaternary Studies* (eds R. E. Leakey and B. A. Ogot): pp. 268–70. Nairobia: TILLMIAP.

_____. and Stemler, A. B. L. 1975. Early domesticated sorghum from central Sudan. *Nature* 254:588–91.

_____., Williams, M. A. J. and Smith, A. B. 1973. The geomorphology and archaeology of Adar Bous, central Sahara: a preliminary report. *Quaternaria* 17:245–97.

Coudé-Gaussen, G. 1984. Le cycle des poussières éoliennes désertiques actuelles et la sédimentation des loess péridésertiques quaternaires. *Bull. Centres Rech. Explor.-Prod. Elf-Aquitaine* 8:167–82.

_____. and Rognon, P. 1983. Les poussières sahariennes. *La Recherche* 147:1050–61.

Douglas, I. and Spencer, T. (eds) 1985. *Environmental Change and Tropical Geomorphology*. London: George Allen and Unwin.

Edmunds, W. M. and Wright, E. P. 1979. Groundwater recharge and palaeoclimate in the Sirte and Kufra basins, Libya. *Journal of Hydrology* 40:215–41.

Gabriel, B. and Kröpelin, S. 1983. Jungquartäre limnische Akkumulationsphasen im NW-Sudan. *Z. Geomorph. NF Suppl.-Bd.* 48:131–43.

_____. 1984. Holocene lake deposits in northwest Sudan. *Palaeoecology of Africa* 16:295–299.

Gasse, F. 1975. *L'évolution des lacs de l'Afar central (Ethiopie et T.F.A.I.) du Plio-Pléistocène à l'actuel*. (Thèse, Docteur ès Sciences Naturelles, l'Université de Paris, 3 vols).

_____. 1977. Evolution of Lake Abhé (Ethiopia and T.F.A.I.) from 70,000 bp. *Nature* 265:42–5.

Grove, A.T. 1958. The ancient erg of Hausaland and similar formations on the south side of the Sahara. *Geogr. J.* 124:528–33.

_____. 1985. The physical evolution of the river basins. In *The Niger and Its Neighbours* (ed A.T. Grove): pp. 21–60. Rotterdam: Balkema.

Harvey, C. P. D. and Grove, A. T. 1982. A prehistoric source of the Nile. *Geographical Journal* 148:327–336.

Hassan, F. 1980. Prehistoric settlements along the Main Nile. In *The Sahara and the Nile* (eds M. A. J. Williams and H. Faure): pp. 421-50. Rotterdam: Balkema.

Jäkel, D. 1979. Run-off and fluvial formation processes in the Tibesti Mountains as indicators of climatic history in the cental Sahara during the late Pleistocene and Holocene. *Palaeoecology of Africa* 11:13–44.

Kendall, R. L. 1969. An ecological history of the Lake Victoria basin. *Ecological Monographs* 39:121–76.

Klitzsch, E., Sonntag, C., Weistroffer, K. and El-Shazly, E. M. 1976. Grundwasser der Zentralsahara: Fossile Vorräte. *Geologische Rundschau* 65:264–87.

Kolla, V., Biscaye, P. E. and Hanley, A. F. 1979. Distribution of quartz in Late Quaternary Atlantic sediments in relation to climate. *Quat. Res.* 11:261–77.

Krzyaniak, L. 1980. The origin of pastoral adaptation in the Nilotic savanna. In *Proceedings of the 8th Panafrican Congress on Prehistory and Quaternary Studies* (eds R. E. Leakey and B. A. Ogot): pp. 265–7. Nairobi: TILLMIAP.

Leakey, R. E. and Ogot, B. A. (eds) 1980. *Proceedings of the 8th Panafrican Congress on Prehistory and Quaternary Studies.* Nairobia: TILLMIAP.

Livingstone, D. A. 1980. Environmental changes in the Nile headwaters. In *The Sahara and the Nile* (eds M. A. J. Williams and H. Faure): pp. 339–59. Rotterdam: Balkema.

Mainguet, M., Canon, L. and Chemin, M. C. 1980. Le Sahara: géomorphologie et paléogéomorphologie éoliennes. In *The Sahara and the Nile* (eds M. A. J. Williams and H. Faure): pp. 17–35. Rotterdam: Balkema.

Messerli, B. and Winiger, M. 1980. The Saharan and East African uplands during the Quaternary. In *The Sahara and the Nile* (eds M. A. J. Williams and H. Faure): pp. 87–118. Rotterdam: Balkema.

Mori, F. 1965. *Tadrart Acacus, Arte Rupestre e Culture del Sahara Preistorico.* Turin: Einaudi.

Munson, P. J. 1975. Archaeological data on the origins of cultivation in the southwestern Sahara and its implications for West Africa. In *Origins of African Plant Domestication* (eds J. R. Harlan, J. M. J. de Wet and A. B. L. Stemler): pp. 187–210. The Hague: Mouton.

Paquet, H., Cloudé-Gaussen, G., Rognon, P. and Wendling, R. 1984. Etude minéralogique de poussières sahariennes le long d'un itinéraire entre 19° et 35° de latitude nord. *Rev. Géol. Dyn. Géogr. Phys.* 26:257–65.

Parkin, D. W. and Shackleton, N. 1973. Trade-winds and temperature correlations down a deep-sea core off the Saharan coast. *Nature* 245:455–7.

Perrott, F. A. 1979. *Late Quaternary lakes in the Ziway-Shala basin, southern Ethiopia.* (Ph.D. Thesis, University of Cambridge).

Quézel, P. and Martinez, C. 1958. Etude palynologique de deux diatomites du Borkou (Territoire du Tchad, AEF). *Bull. Soc. Hist. Nat. Afr. Nord.* 49: 230–44.

_____. 1962. Premiers résultats de l'analyse palynologique de sédiments recueillis au Sahara méridional à l'occasion de la

Mission Berliet. Tchad. In *Mission Berliet Ténéré—Tchad* (ed H. J. Hugot): pp. 313–27. Paris.

Ritchie, J. C., Eyles, C. H. and Haynes, C. V. 1985. Sediment and pollen evidence for an early to mid-Holocene humid period in the eastern Sahara. *Nature* 314:352–55.

Rossignol-Strick, M., Nesteroff, W., Olive, P. and Vergnaud-Grazzini, C. 1982. After the deluge: Mediterranean stagnation and sapropel formation. *Nature* 295:105–10.

Sarnthein, M. 1978. Sand deserts during glacial maximum and climatic optimum. *Nature* 272:43–6.

———, Tetzlaff, F., Koopman, B., Wolter, K. and Pflaumann, U. 1981. Glacial and interglacial wind regimes over the eastern subtropical Atlantic and North-West Africa. *Nature* 293:193–6.

Servant, M. and Servant-Vildary, S. 1980. L'environnement quaternaire du bassin du Tchad. In *The Sahara and the Nile* (eds M. A. J. Williams and H. Faure): pp. 133–62. Rotterdam: Balkema.

Sonntag, C., Thorweihe, U., Rudolph, J., Löhnert, E. P., Junghaus, C., Münnich, K. O., Klitzsch, E., El-Shazly, E. M. and Swailem, F. M. 1980. Isotopic identification of Saharan groundwaters, groundwater formation in the past. *Palaeoecology of Africa* 12:159–71.

Stanley, D. J. and Maldonado, A. 1977. Nile cone: Late Quaternary stratigraphy and sediment dispersal. *Nature* 266:129–35.

Stemler, A. B. L. 1980. Origins of plant domestication in the Sahara and the Nile Valley. In *The Sahara and the Nile* (eds M. A. J. Williams and H. Faure): pp. 503–26. Rotterdam: Balkema.

Street, F. A. and Grove, A. T. 1976. Environmental and climatic implications of late Quaternary lake level fluctuations in Africa. *Nature* 261:385–90.

Street-Perrott, F. A., Roberts, N. and Metcalfe, S. 1985. Geomorphic implications of late Quaternary hydrological and climatic changes in the Northern Hemisphere tropics. In *Environmental Change and Tropical Geomorphology* (eds I. Douglas and T. Spencer): pp. 165–83. London: George Allen and Unwin.

Sutton, J. E. G. 1977. The African aqualithic. *Antiquity* 51:25–34.

Talbot, M. R. 1980. Environmental responses to climatic change in the West African Sahel over the past 20,000 years. In *The Sahara and the Nile* (eds M. A. J. Williams and H. Faure): pp. 37–62. Rotterdam: Balkema.

———, and Delibrias, F. 1980. A new Late Pleistocene—Holocene water-level curve for Lake Bosumtwi, Ghana. *Earth Planet. Sci. Lett.* 47:336–44.

Tothill, J. D. 1946. The origins of the Sudan Gezira clay plain. *Sudan Notes and Records* 27:153–83.

Vita-Finzi, C. 1973. *Recent Earth History*. London: Macmillan.

Wendorf, F. and Schild, R. 1980. *Prehistory of the Eastern Sahara*. New York: Academic Press.

———, and Close, A. E. (eds) 1984. *Cattle-Keepers of the Eastern Sahara: the Neolithic of Bir Kiseiba*. Dallas: SMU Press.

Williams, M. A. J. and Faure, H. (eds) 1980. *The Sahara and the Nile*. Rotterdam: Balkema.

———. 1984. Late Quaternary prehistoric environments in the Sahara. In *From Hunters to Farmers: The Causes and Consequences of Food Production in Africa* (eds J. D. Clark and S. A. Brandt): pp. 74–83. Berkeley and Los Angeles: University of California Press.

———. 1985a. Pleistocene aridity in tropical Africa, Australia and Asia. In *Environmental Change and Tropical Geomorphology* (eds I. Douglas and T. Spencer): pp. 219–33. London: George Allen and Unwin.

———. 1985b. On becoming human: geographical background to cultural evolution. *Australian Geographer* 16:175-84.

———, and Adamson, D. A. 1980. Late Quaternary depositional history of the Blue and White Nile rivers in central Sudan. In *The Sahara and the Nile* (eds M. A. J. Williams and H. Faure): pp. 281–304. Rotterdam: Balkema.

———, and Adamson, D. A. (eds) 1982. *A Land between Two Niles: Quaternary Geology and Biology of the Central Sudan*. Rotterdam: Balkema.

———, Adamson, D. A. and Baxter, J. T. 1986. Late Quaternary environments in the Nile and Darling basins. *Aust. Geog. Studies* 24:28-44.

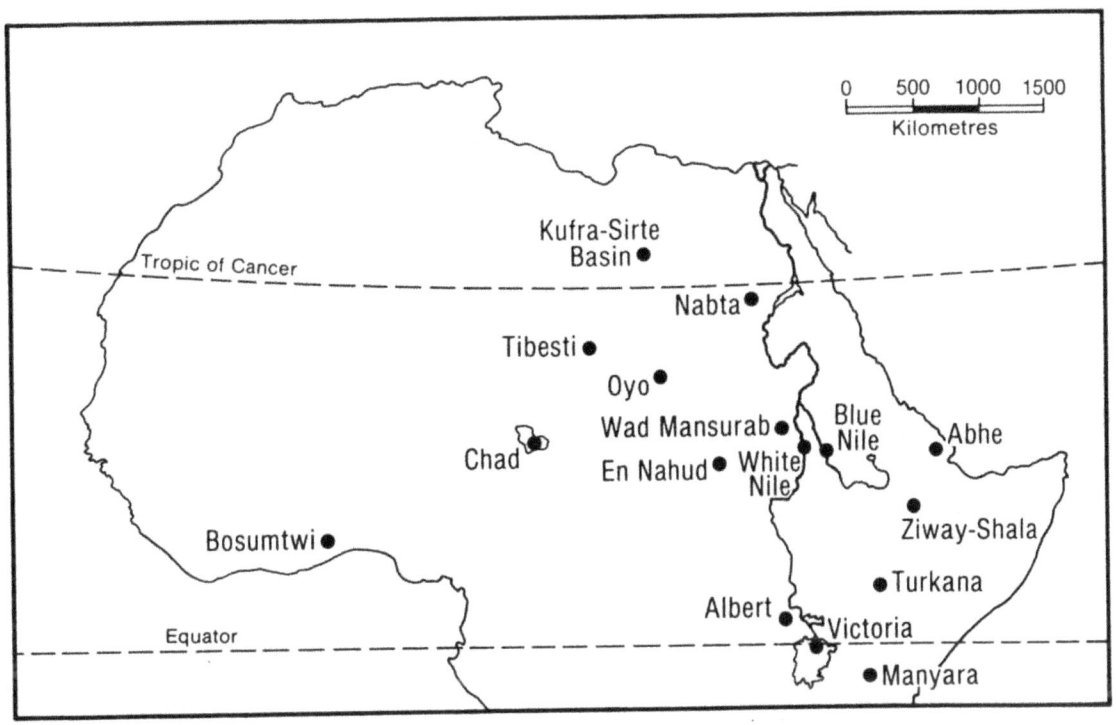

Fig. 3.1: Location of sites illustrated in Fig. 3.2 and 3.3.

After the Deluge: The Neolithic Landscape in North Africa

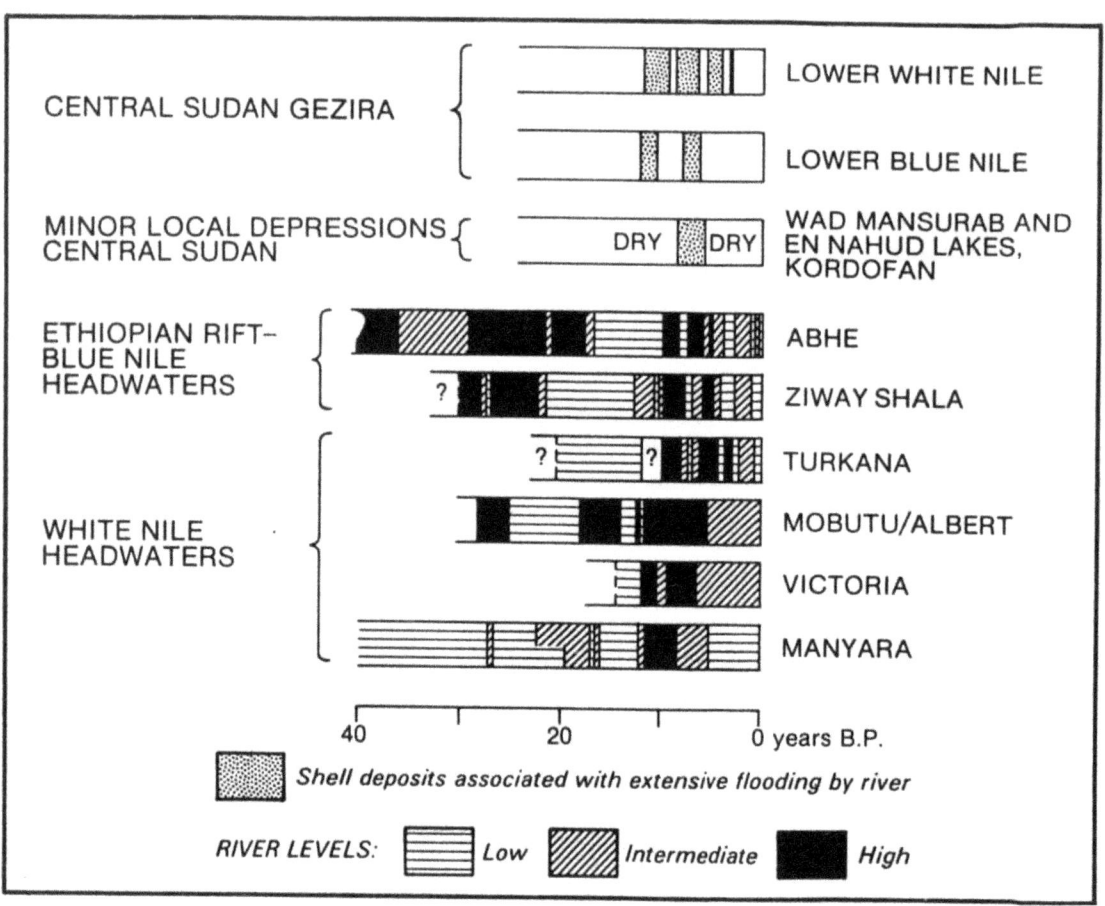

Fig. 3.2: Times of high late Quaternary flood levels in the lower Blue and White Nile valleys, central Sudan, and corresponding fluctuations in lake levels in and around the Nile basin (after Adamson et al. 1980; Williams and Adamson 1980; and Adamson 1982).

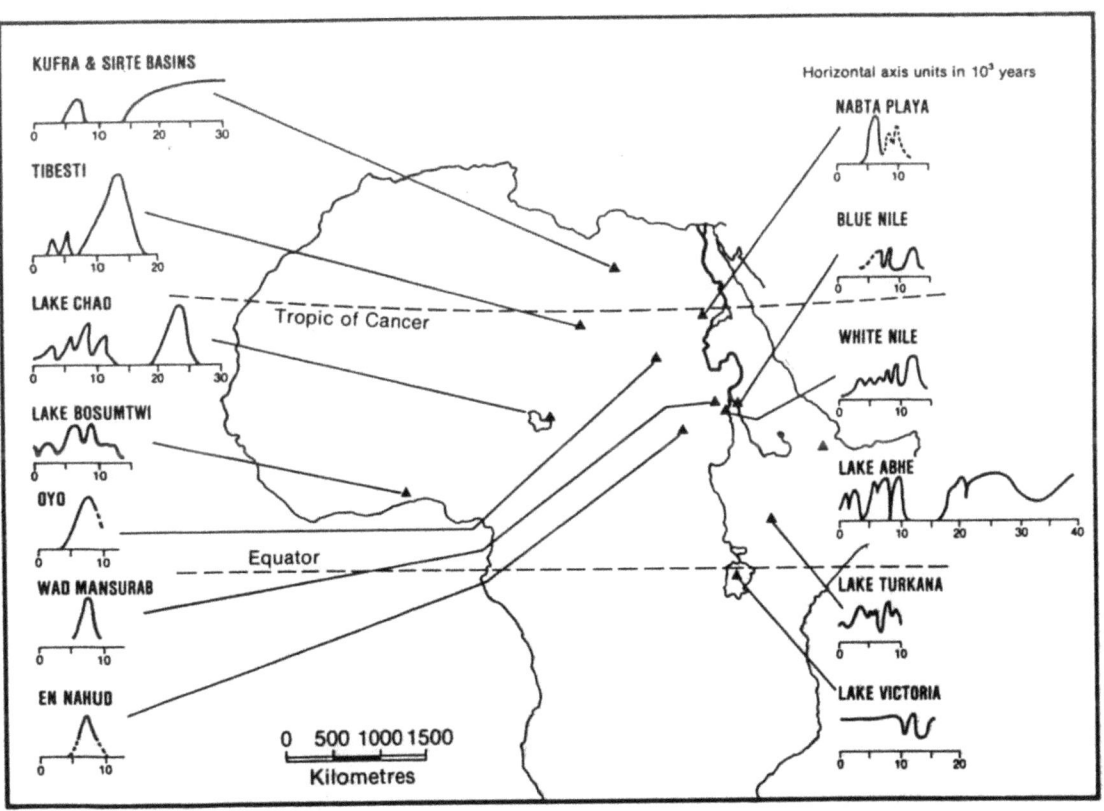

Fig. 3.3: Late Quaternary hydrological events at selected localities in North Africa. The peaks denote high lake levels, high flood levels in the lower Blue and White Nile valleys, greater runoff from the Tibesti uplands, and increased groundwater recharge in the Kufra and Sirte basins of central Libya (compiled from date in Kendall 1969; Klitsch *et al.* 1976; Gasse 1977; Edmunds and Wright 1979; Jäkel 1979; Adamson *et al.* 1980; Butzer 1980; Servant and Servant-Vildary 1980; Sonntag *et al.* 1980; Wendorf and Schild 1980; Adamson 1982; Adamson *et al.* 1982; Williams 1984, 1985a; Ritchie *et al.* 1985; and Williams et al. 1986).

After the Deluge: The Neolithic Landscape in North Africa

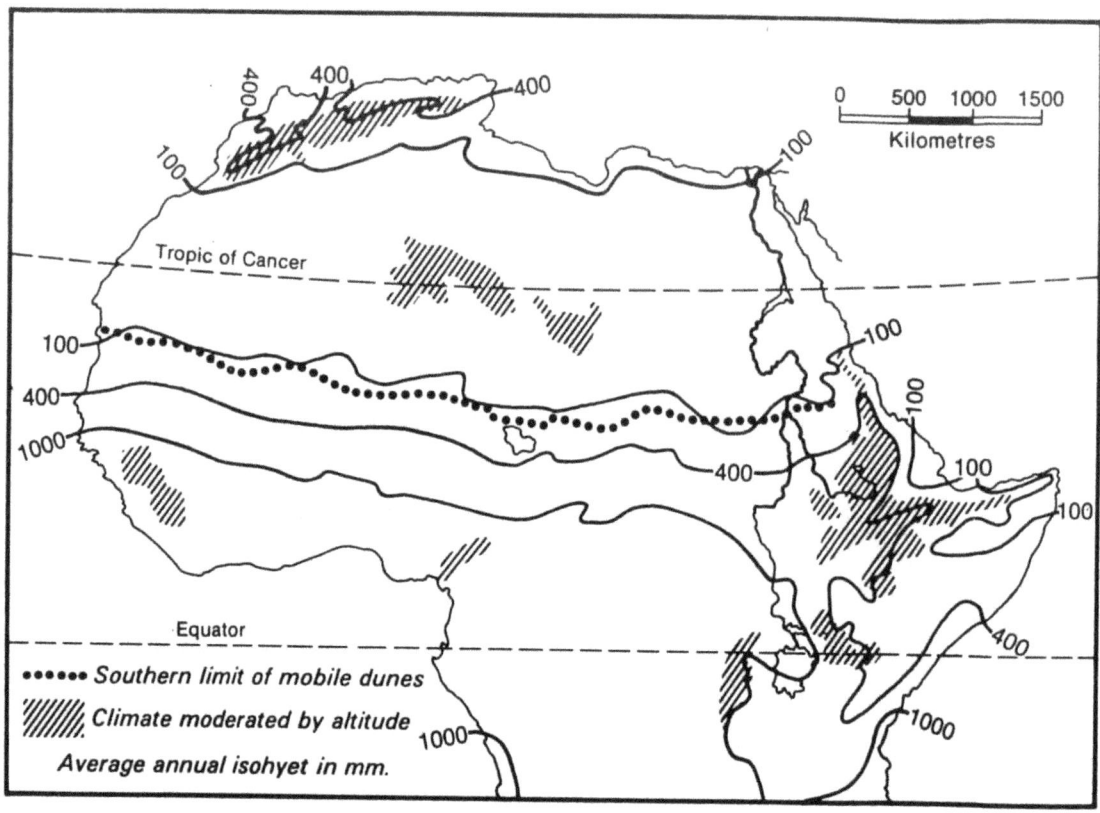

Fig. 3.4: Present-day rainfall zones in North Africa (from *Atlas of Africa* 1973). The southern limit of presently mobile dunes coincides approximately with the 100mm isohyet (from Mainguet *et al.* 1980).

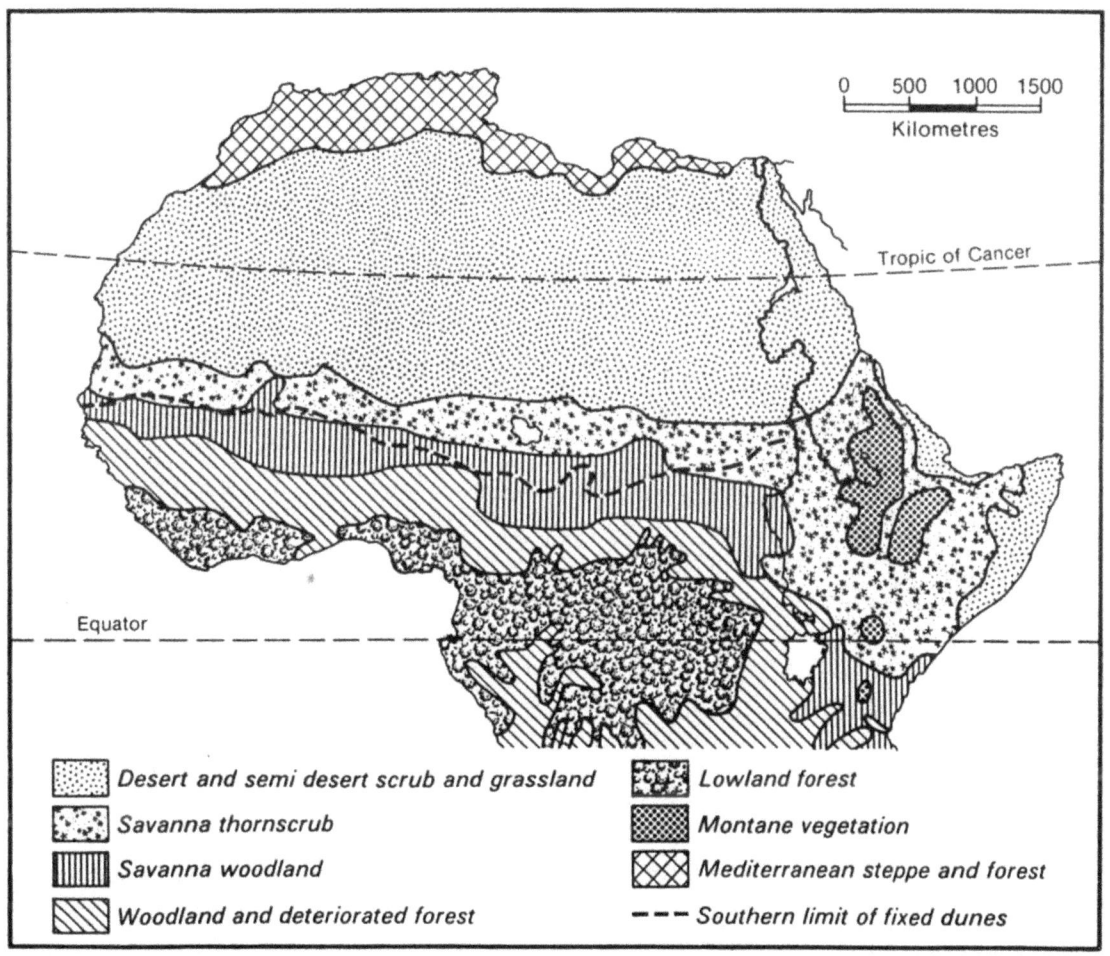

Fig. 3.5: Present-day vegetation zones in North Africa (from *Atlas of Africa* 1973). The belt of fixed late Pleistocene dunes lies in the present-day savannah thornscrub and savanna woodland zones of the Sahel and the Sudan (from Mainguet *et al.* 1980).

After the Deluge: The Neolithic Landscape in North Africa

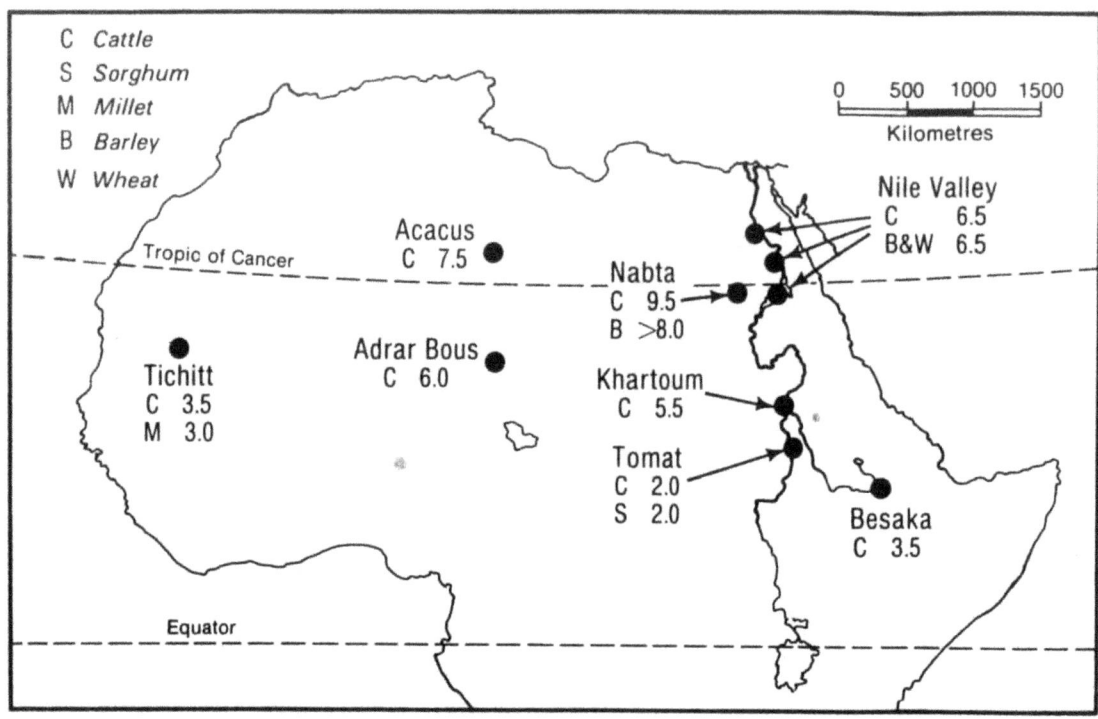

Fig. 3.6: First recorded appearance of certain domesticated plants and animals at selected sites in North Africa (after Williams 1985b).

Neolithic Adaptations on the Central Nile

By Abbas Mohammed-Ali, Department of Archaeology, University of Khartoum, and El-Sayed El-Anwar, Department of Archaeology, University of Bergen, Norway.

As a result of recent archaeological investigations along the east bank of the Nile in Khartoum Province, Sudan (Fig. 4.1), a hypothesis was developed to explain Neolithic adaptations along the Central Nile (Haaland 1978, 1981a, 1981b). In the area surveyed, four early ceramic sites (Kadero II, Um Direiwa I, Um Direiwa II, and Zakiab I) (Fig. 4.1) were located and were test excavated. A fifth site (Kadero I), which was then undergoing excavations (Krzyzaniak 1975), was also considered when the model was formulated, as were the results from a brief survey well back from the Nile.

It was observed that four of these sites (Kadero I and II, Um Direiwa I and II), shared certain features. They are large, occupying areas of between 10,000 m^2 and 30,000 m^2 each, and they are situated, on average, about 7 km from the present river. Excavations revealed abundant quantities of pottery and implements associated with pottery production such as burnishers, red ochre, and serrated shells, which were thought to reflect on-site pottery manufacturing. The sites were also rich in grinding implements and lithic artifacts and had burials associated with them. The fifth site, Zakiab I, is a small settlement, only 2000 m^2 in area, and it is located only 3 km from the present river. It was rich in lithic debitage and had numerous fish bones, while fishing tools such as fish hooks were also present. Pottery manufacturing implements and grinding tools were less frequent than at the other sites (Haaland 1981; El-Mahi 1981). Looking for rainy season herding camps, Haaland (1981) also conducted a survey farther inland, to the east, but this resulted in no significant finds.

On the basis of the distribution of these sites and their cultural manifestations, it was postulated that they reflected a settlement pattern related to seasonally specific activities. The four large sites were seen as permanent base camps where emphasis was placed upon the exploitation of plants, sorghum cultivation, and the manufacturing of pottery. The small site (Zakiab I) was interpreted as a dry season camp where herding and fishing were practiced. It was temporarily occupied and was also seen as a place where lithic artifacts were manufactured. The model proposed that a large community occupied a base camp when conditions were favourable for cultivation. During the dry season, the inhabitants of each base camp would split into smaller bands and occupy fishing and herding camps along the Nile where conditions would be optimal for these activities. After the rains, equivalent herding camps would be set up in the grasslands of the Butana farther to the east.

Haaland's model, though based on limited data, seems reasonable. We thought of testing it within the same general area but with sites found on the west bank of the Nile (Fig. 4.1).

In a survey conducted along the west bank of the Nile (Mohammed-Ali 1984), a number of early ceramic sites were located, all of which were confined to a 1 km strip along the river. In spite of systematic survey, no sites were located in the 5 km strip farther to the west. Two of the sites, NOF I, located at the village of Nofalab, and ISG I, situated at Islang village, both about 25 km north of Omdurman (Fig. 4.1) were selected for excavations (El-Anwar 1981, 1982). The reason behind this selection was that one of them, NOF, I was comparable in size to Kadero I (Krzyzaniak, 1975) and Um Direiwa I and II (Haaland, 1981a), large sites which would be classified as base camps under Haaland's model, while the other, ISG I, was small and according to Haaland's criteria might have been thought of as a dry season camp. The cultural material on the surface of both sites was seen to be contemporary; that is, the ceramics were typical of those falling within the Khartoum Neolithic (Arkell 1953).

The NOF I site (lat. 15–59 N., long. 32–32 E.) is located on the west bank of the Nile River about 23 km north of Omdurman (Fig. 4.1). It is situated on a gravel ridge which is believed to be an old terrace of the Nile (Arkell 1953:1; El-Anwar 1982:9–11). Because of the meandering course of the river, this gravel ridge is, at present, 750 m away from the river. The elevation of NOF I is about 7 m above the present flood plain.

On the basis of the distribution of the material on the surface and in the excavated test pits, it is estimated that NOF I occupies an area of 180 m x 170 m (*ca.* 30,600 m^2), making it one of the larger Khartoum Neolithic sites. The excavations at NOF I revealed variable depths of the cultural deposit, between 40 cm and 80 cm. A wide range of material culture was recovered, including abundant lithic artifacts, ceramics, and floral and faunal remains. However, no evidence was found for an earlier Khartoum Mesolithic occupation.

The result of the study made on the lithic artifacts shows utilization of raw materials which were either obtained locally or brought from farther north. Being locally available, quartz was the most frequently used raw material in tool production, constituting 96.3% of all artifacts. but only 62.7% of the tools. Petrified wood and sandstone were used for hammer stones and grinding implements. Rhyolite, basalt, and other volcanic rocks unavailable in the vicinity of the site were imported from the Sabaloca Gorge some 47 km to the north (Fig. 4.1) and were used for chipped stone tools.

The technology is dominated by the production of microlithic flakes. Whether the production of microliths was a cultural preference or was determined by the size of the locally available quartz pebbles is difficult to know. The occurrence of rhyolite debitage at the site favours the former interpretation. Large nodules of rhyolite must have been brought into the site and used locally for the manufacture of larger implements.

The ceramics from NOF I resemble those recorded by Arkell (1953) from Shaheinab. They are thin-walled and made of fine-grained clay with quartz inclusions. No complete pots were found to allow for the determination of vessel forms, and no implements were recovered that could be interpreted as evidence for pottery manufacture. The majority of the sherds are burnished and on some of them a red slip was applied. Among the decorated sherds, the techniques employed were incising and rocker stamping. The motifs include dotted zigzag, continuous straight lines, and bands of triangles, and "vees." The latter exhibit considerable variation but are always well made.

The macrobotanical remains attest to the presence of a number of taxa but show no evidence of any domestic plants which might have been utilized. Seeds of *Zizyphus* sp. and abundant seeds of *Celtis integrifolia*, Lam. were found. Sherds from this site were carefully examined for plant impressions. Only one seed impression of *Celtis integrifolia* Lam. and one seed impression of *Sorghum verticilliflorum* were identified.

The identified osteological remains from NOF I site exhibit a wide variety of riverine and terrestrial species. These include: shells of *Pila wernei*, *Lanistes carinatus* and *Limicolaria cailliaudi* which are very frequent; *Cleopatra bulimoides* and *Eitheria elliptica* are also present (Al-Mahi, personal communication). Aquatic mammals and fish remains of different species were found in considerable abundance, *Clarias* sp. (catfish) being the most numerous; remains of *Hippopotamus amphibius* were also found. The terrestrial species include a variety of wild forms such as antelope, as well as domestic animals such as cattle (*Bos taurus*), sheep (*Ovis aries*), and goat (*Capra hircus*). In fact, the majority of the identified bones are from domestic animals.

The Neolithic site of Islang (Lat. 15–53 N., Long. 32–32E.) is located on a crest of an eroded sandstone ridge rising about 9 m above the surrounding plain and about 1 km north of the village of Islang and 25 km north of Omdurman (Fig. 4.1). It lies 0.85 km from the present west bank of the Nile.

Neolithic Adaptations on the Central Nile

The site was badly disturbed by Meroitic burials which cut through its stratigraphy but enough was left to allow test excavations into primary context deposits. It was estimated to have originally occupied 68 m x 66 m (*ca.* 4488 m²) comparable to a seasonal camp under Haaland's model. ISG I revealed two cultural components that are stratigraphically positioned. The lower component produced a "Khartoum Mesolithic" assemblage with wavy-line pottery, while the upper component is similar to that recovered by Arkell (1953) at Shaheinab, widely known as "Khartoum Neolithic." Our concern here lies with the latter.

The Neolithic cultural deposits at ISG I vary from 20 to 30 cm in depth. The data recovered from the site included lithic artifacts, potsherds, and faunal and floral remains.

The raw materials utilized among the lithic artifacts are mostly available in the vicinity of the site. Quartz is by far the most common among the flaked artifacts; rhyolite and basalt were also used. Petrified wood and sandstone were rather favoured among pounding/grinding implements. Apart from rhyolite and basalt, which are believed to have been brought from the Sabaloca Gorge of the Sixth Cataract (Fig. 4.1), all the other rock types are of local origin. Quartz counts for 92% of the raw materials used at ISG I but it drops to 67% among the tools, giving way to rhyolite and basalt.

The ceramic component from ISG I also recalls that of Shaheinab. It is dominated by fabrics with fine-grained quartz inclusions and by a high percentage of burnished sherds which are sometimes slipped. Here, combing, incising, and rocker stamping were the main techniques employed, while triangles were the most common motif.

One badly fragmented seed of *Celtis integrifolia* Lam. was recovered. When the sherds were examined for plant impressions, no evidence was found.

A preliminary study of the faunal remains from the Islang site attests to the presence of different species, many of which are associated with riverine habitats. These include gastropods such as *Pila wereni*, and *Limicoloria cailliaudi*, different species of fish such as *Clarias sp.* and *Lates niloticus*, together with reptiles such as *Crocodylus niloticus* and *Varanus niloticus*. All these provide clear evidence for riverine exploitation. The identification of the mammalian fauna is not yet complete, but domestic animals are expected.

The available floral and faunal evidence from the NOF I site indicates that wetter climatic conditions prevailed during the Neolithic occupation compared with the current arid conditions. The presence of abundant seeds of *Celtis integrifolia* at the site supports this interpretation, since this species requires a minimum rainfall of 400–500 mm. In addition, the presence of swamp and semi-aquatic gastropod species at the site, such as *Cleopatra bulimoides*, *Lanistes carinatus*, *Pila wernei*, and *Limicolaria cailliaudi* provide further evidence for wetter climate than that of the present day.

The faunal and floral remains found at NOF I indicate a type of mixed subsistence economy based on domesticated animals (cattle, sheep, and goats); fishing was clearly one of the major food quest activities, while hunting of wild game, such as antelopes, seemed to have been a supplementary source of food. The exploitation of plants is also in evidence as the presence of seeds of *Zizyphus* sp. and *Celtis integrifolia* indicate the collection of their edible fruits.

The fragmented seed of *C. integrifolia* found in ISG I may be too meager evidence by itself to allow a reconstruction of the climatic and/or environmental conditions during the occupation of the site. Yet, among the abundant faunal material that was recovered, some was associated with a riverine environment. Since this site is at least partly contemporary with NOF I, it may, however, be possible to postulate a tentative climatic reconstruction.

The results obtained from the analysis of the floral and faunal remains so far indicate that the occupants of ISG I might have kept domesticated animals, but also practiced a riverine adaptation based on the exploitation of aquatic resources. However, unlike the sites on the east bank, comparison of materials from ISG I and NOF I neither shows clear differences in activities nor reflects evidence of specific seasonal adaptations.

The radiocarbon dates obtained from NOF I (5290 bp ±100, T-3700; 5520 bp ±130, T-3701) are close to those obtained from Kadero I (5500 bp ±70, KN-2821; 5260 bp ±90, T-2188; and Zakiab, 5660 bp ±80, T-3050). No radiocarbon date was obtained from the Neolithic component at ISG I but a date of 5870 bp ±110 (T-3880) was obtained from the Mesolithic level just below it. This would give the Neolithic component of ISG I a post-5800 bp date. The dates from Kadero II, Um Direiwa I and II (Haaland 1981:55–6) also cluster between 5500–5000 bp. The ceramic assemblages from all of those sites exhibit close similarities within the Khartoum Neolithic tradition.

There are reasons to believe that on the Nile the Khartoum Neolithic was a short-lived tradition and that all sites representing it are at least partly contemporary (Mohammed-Ali 1987). Thus, the variations in faunal and artifact site inventories cannot reflect diachronic change but must relate to the different behavioral patterns which might either reflect seasonal differences or microenvironmental adaptations. Our evidence suggests that the latter may have greater significance than realized to date.

A clue for this may lie in the fact that the river divides the area topographically into two distinct regions. The east bank which provided the data for Haaland's model is a flat steppe which marks the western edge of the Butana plain. It is covered by Butana clays and Gezeira alluvial deposits. The western bank, where our sites are located, consists, apart from a narrow strip of flood plain, of sandstone formations and other lithological outcrops. Considering these topographic and geological differences, it may well be that each side of the river saw different patterns of settlement.

While the sites on the east bank are scattered in an area extending 7 km inland from the river, those on the west bank are close to the river. The settlement pattern on the west bank does not suggest occupation back from the river as is the case of the east bank. It is expected that the topographical differences between the two banks must have affected local adaptations. The flat alluvial clays of the east bank with their Nile-fed swamps and ponds would allow cultivation to be practiced and would support a rich pasture with a thick cover of vegetation and shrubs. On the other hand, the eroded sandstone and pebble conglomerates of the west bank would not permit agriculture, and their stony surfaces support relatively little grass even after the rainy season. If the Nile waters were at least 3–5 m higher during the Neolithic than their present level (Arkell 1953), it would mean that the narrow strip of the flood plain, currently used for cultivation, would have been under water.

At least on the west bank the sites seem to have been occupied permanently rather than seasonally, suggesting that the rich and varied riverine environment was capable of providing stable sources of subsistence for the inhabitants. Our two sites reveal similar percentages of grinding stones and scrapers, and they do not show any clear differences in the utilization of fish or domestic animals. The fact that their sizes are different does not seem to reflect different adaptation or even different on-site activity patterns. Both seem to reflect a wide range of activities which are spatially discrete on the east side of the Nile.

Taking the situation as it stands on the west bank and the inapplicability of Haaland's model, we thought of a reversed pattern to that proposed by Haaland. This model assumes that, since the large sites are located close to the river, these sites might have served as base camps, densely populated during the dry season. When conditions improved in the hinterlands, during the rainy season, part of the population might have split into small groups and occupied smaller sites in those areas. The evidence available from the region today does not yet confirm this. However, the question should be left open, awaiting surveys in the areas even further west. Should these surveys turn up ephemeral inland sites, then it might be indicated that two quite different settlement systems existed on the opposite banks of the Nile River during the Khartoum Neolithic.

References

Arkell, A. J. 1953. *Shaheinab*. London: Oxford University Press.

El-Anwar, E. 1981. Archaeological excavations on the west bank of the river Nile in the Khartoum area. *Nyame Akuma* 18:43–45.

_____. 1982. *The Khartoum Neolithic in the light of archaeoethnobotany: A case study from the Nofalab and the Islang sites.* (Unpublished M.A. thesis, University of Khartoum).

El-Mahi, A. T. 1981. The interpretation of the earliest evidence of lungfish *Protopterus* sp. in the Nile Valley-El-Zakiab site, Central Sudan. *Norwegian Archaeological Review* 14(1):60–65.

Haaland, R. 1978. Ethnographical observation of pottery-making in Darfur, Western Sudan, with some reflection on archaeological interpretations. In *New Directions in Scandinavian Archaeology* (eds K. Kristiansen and P. Muller): pp. 47–61. Copenhagen.

_____. 1981a. *Migratory herdsmen and cultivating women: The structure of Neolithic seasonal adaptation in the Khartoum Nile environment.* Bergen.

_____. 1981b. Seasonality and division of labor: A case study from Neolithic sites in the Khartoum Nile environment. *Norwegian Archaeological Review* 14(1):44–50.

Krzyzaniak, L. 1975. Polish excavations at Kadero. *Nyame Akuma* 7:45–6.

Mohammed-Ali, A. 1984. Sorourab-I: A Neolithic site in Khartoum Province, Sudan. *Current Anthropology* 25 (1):117–19.

Mohammed-Ali, A. 1987. The Neolithic of Central Sudan: a reconsideration. In *Prehistory of Arid North Africa* (ed A. Close): pp. 123-36. Dallas: Southern Methodist University Press.

Neolithic Adaptations on the Central Nile

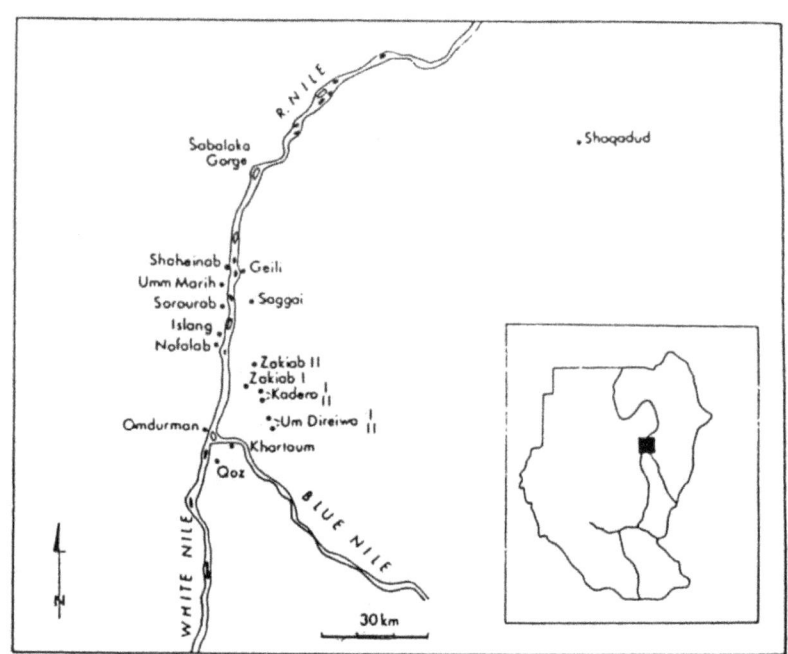

Fig. 4.1: Neolithic sites in the Central Nile.

Holocene Environments and Occupations in the Southern Atbai, Sudan: A Preliminary Formulation

By Anthony E. Marks and Karim Sadr, Southern Methodist University, Dallas, Texas.

Introduction

Since 1979, intensive archaeological surveys and test excavations have been in progress in the far eastern Sahel, in the area between the Ethiopian border and the Atbara River (Fig. 5.1). This work has been carried out by two archaeological projects: the Joint University of Khartoum/Southern Methodist University Butana Archaeological Project (Marks et al. 1980, 1982; Marks 1984; Mohammed-Ali and Marks 1984) and the Italian Mission to Kassala (Coltorti et al. 1984; Constantini et al. 1982, 1983; Durante, Fattovich and Piperno 1980; Fattovich and Piperno 1981). Although field work is now going on and probably will continue in the years to come, it is possible to present a tentative formulation of the relationships between changing settlement patterns, changes in regional and local environmental conditions, and apparent shifts in economic adaptations. The temporal scope of this paper will be limited to the Holocene from about 7000 BC until sometime during the first millennium AD. The occupation which postdates the first millennium AD, called the Gergaf Group, has already seen preliminary publication (Sadr 1984) and need not be repeated here.

During the past decade, a number or studies have been made which throw light on past climatic changes in the Central Sudan (e.g., Grove 1973; Grove and Goudie 1971; Warren 1970; Wickens 1975, 1982; Williams 1974). Although there are minor discrepancies in the details of their conclusions, there is general agreement that a few, broad climatic periods can be recognized. In the broadest terms, these periods document a slow drying trend from an Early Holocene base which was much wetter than today (Grove 1973), through an intermediate period during the Middle Holocene (*ca.* 6000 bp to 3000 bp) when climatic conditions were wetter than today and vegetational belts were some 100 km north of their present position (Warren 1970), to a final period with a climatic regime like that of the present day (Wickens 1975, 1982).

Today, the Southern Atbai receives somewhat over 300 mm of annual precipitation which falls during the summer months (Barbour 1964). The Southern Atbai is classified vegetationally as a thorn savanna and scrub region. Since the soils are mainly clay, the major vegetation is *Acacia mellifera* with a number of annual grasses (Wickens 1982). Under early Holocene conditions, the area has been postulated to have been covered with a deciduous savanna woodland, the major trees of which would have been *Combretum hartmannianum* and *Anogeissus leiocarpus*. This type of savanna is present today along the Ethiopian border some 175 km to the south of Kassala (Wickens 1975:52). During the Middle Holocene it appears that no great change took place in the vegetation of the Southern Atbai, although there was an apparent drop in precipitation (Wickens 1975). The major change in environment, that which brought about a significant shift in floral patterns, took place only after 3000 bp.

In this regard, the regional climatic history of the Southern Atbai provides only the flimsiest of backdrops onto which the changing patterns of human settlement and adaptation may be projected. These regional reconstructions, by their very nature, cannot deal realistically with local conditions; those affected areas like the Southern Atbai cover perhaps no more than a few tens of thousands of square kilometers. It is these latter, local environmental conditions and their fluctuations which, in this case, appear to have had major impact on the history of Holocene settlement patterns.

In order to understand what happened in the Southern Atbai during the Holocene, it is necessary to realize that a major portion of its moisture comes from rivers, the waters of which originate well to the south, outside the climatic zone of the Central Sudan.

Two rivers partly define the area: the Atbara River forms the western border of the Southern Atbai and the Gash River flows close to its eastern edge (Fig. 5.1). The Atbara derives most of its water from monsoonal rains in the area of Lake Tana in Ethiopia. The river is deeply incised into the steppe through which it runs and, although it has a very high seasonal load during the rainy season, this has little effect on its surrounding terrain. A quite different situation pertains to the Gash. Its catchment is some 21,000 km^2 of Ethiopian hill slopes which, even today, receive over 500 mm of rain during the summer months (Barbour 1964:219). From the Gash River, the Southern Atbai receives massive floods between June and September. Today, and probably since the Middle Holocene, this water has no outlet, forming instead an inland delta which has been extensively developed for agriculture. In fact, most of the eastern half of the Southern Atbai is made up of ancient deltaic deposits of the Gash River (Durante *et al.* 1980).

In addition to the riverine hydrology, some 320 mm of rain falls each summer in the upper Gash Delta, with only a bit less falling to the west near the Atbara. Thus, even today, the Southern Atbai has considerable water, too much, in fact, during the summer months, with the greatest amount on the eastern edge, along the Gash, and the least on the Western edge, just east of the deeply cut Atbara valley.

It is quite clear that the Holocene geomorphic history of the Southern Atbai has been one of a slow northeasterly movement of the Gash Delta (Durante *et al.* 1980). It appears that during the Terminal Pleistocene or, more likely, the Early Holocene, the Gash flowed into the Atbara a bit north of Khashm el Girba (Barbour 1964). Having a very shallow bed, however, and alternating from dry to massive flooding, the Gash has been characterized by much meandering and enormous lateral overbank inundation. This regime has also brought down from Ethiopia large quantities of sands and silts which are more permeable than the Pleistocene clays of the western half of the Southern Atbai (Fig. 5.2). This has the effect of retaining the seasonal flood waters in the eastern part of the Southern Atbai, while the western half tends to see rapid evaporation of the local precipitation, as well as heavy erosion within the Atbara valley.

Projecting this situation back into the Early Holocene with a different main channel for the Gash (Fig. 5.3), it is clear that seasonal inundation would have reached far to the west, almost to the Atbara. The increased inclination of the river as it flowed into the Atbara valley, however, would have resulted in downcutting and the prevention of significant overbank flooding. As the Gash Delta silted up and moved toward its present north-northwesterly course, the area of seasonal inundation would have moved with it and local induncation between the Atbara and the Gash would have depended more on the amount of precipitation and the degree of soil permeability than on overbank flooding of the Gash. While these changes would not have had major regional impact, they appear to have had significant local effects. It is these local changes, affecting hundreds of square kilometers, which help to explain the changing Holocene settlement patterns in the Southern Atbai.

Cultural Sequence

During most of the Holocene, the Southern Atbai was occupied by various human groups. While these cannot be tied directly to any archaeological cultures already defined for the adjacent areas of the Nile valley (Adams 1977) or the Ethiopian Plateau (Fattovich 1982), it has been possible to construct a local culture historic sequence based mainly on a combination of comparative ceramic assemblages and a series of radiocarbon dates (Fattovich, Marks, and Mohammed-Ali 1984; Marks 1984; Marks and Fattovich, in press). This sequence recognizes both phases and geographic facies of a single ceramic tradition, the Atbai Tradition, which spans much of the Holocene. For the purposes of this chapter, however, usually only the phases need be considered (Fig. 5.4), since the concern is with diachronic change and broad adaptive and settlement patterns, rather than with local stylistic variability.

The earliest Holocene occupation is limited to the Atbara River edge and consists of pre-ceramic, Late Paleolithic hunters who exploited larger mammals such as hippo, giraffe, and wild cattle. Throughout the Early Holocene, these groups were present and it is only with the beginning of the Middle Holocene that ceramic-bearing groups appear. Although it has yet to be documented fully, it seems that the whole of the Southern Atbai ceramic-bearing occupation belongs within a single cultural tradition and that the phases seen are, one way or another, essentially developmental, as well as being sequential. The earliest ceramic occupation is the Pre-Saroba, characterized by ceramics similar to those of the Khartoum Mesolithic in the Nile valley of the Central Sudan (Arkell 1949). However, the similarities are only general and there are elements totally lacking in the Nile valley (Fattovich, Marks, and Mohammed-Ali 1984). The second phase, the Saroba, still exhibits traits similar to those of the Khartoum Mesolithic but, again, they tend to be general rather than specific. The seeming similarities between the Southern Atbai materials and the Khartoum Complex is reinforced by general contemporaneity (Fig. 5.4).

The Saroba Phase passes into the Kassala Phase through a transitional phase which would date in the Nile valley to the very end of the Khartoum Neolithic, Fig. 5.4). Beginning with this transition, what few similarities existed in the material cultures of the Nile valley and the Southern Atbai disappear. From the fourth millennium BC on, the Southern Atbai saw a rich cultural development quite distinct from that in any portion of the Nile valley. The Kassala Phase, particularly its early part, saw the rise of a complex and rich ceramic inventory, the production of and trade for exotic luxury items such as basalts from the Red Sea Hills, and the use of a wide range of ground and polished stone artifacts. Even the chipped stone assemblages are unlike any others in that they remain rich and varied at such a late date.

The Kassala Phase, as now recognized, lasts a long time, from about the beginning of the fourth millennium BC until the beginning of the first millennium BC (Table 5.1). During this phase there is a slow shift in ceramic production from a wide range of decorated wares to a preponderance of simple ceramics and a paucity of fine, decorated wares. In addition, there appears to be a lessening of the amount of luxury items found, although some new forms, such as stone bracelets and disc-shaped mace heads, replace older forms such as pointed mace heads and polished axes. In addition, the later Kassala Phase (post-2000 BC) seems to include a wider range of cultural variability, in that more than two geographic facies can be defined (Fattovich, Marks, and

Mohammed-Ali 1984). By the beginning of the first millennium BC, however, this rich variability seems to have disappeared, leaving a single rather mundane cultural expression during the Jebel Taka Phase. This phase contains sites which show connections, slight though they be, with pre-Axumite materials from Ethiopia and, perhaps, even with the Meroitic on the Nile (Fattovich in press a). Because of these elements, mainly ceramic, the phase has been dated accordingly and the assemblages still await independent confirmation of their age. The Jebel Taka Phase may well last into the first millennium AD when it seems as if the Southern Atbai was either abandoned or the population fell so low that we were unable to find many assemblages which might fit between the Jebel Taka Phase and the near historic Gergaf Group. The only evidence for an occupation during this overall hiatus comes from a few burials in the extreme northeast corner of the survey zone, around the base of Jebel Kamala (Fattovich in press b).

Table 5.1: Radiocarbon Dates from the Southern Atbai

Phase	Group and Site	Date bp	MASCA Calibrated	Lab Number
Late Kassala	Butana (KG96A)	2755 ± 107	984 BC	SMU 1187
	Jebel Mokram (KG20)	3050 ± 90	1350 BC	TX 446
	Gash Group (K1 II)	3860 ± 60	2180 BC	n.d.
Early Kassala	Butana (KG7)	4421 ± 93	3163 BC	SMU 1156
	Butana (KG7)	4569 ± 68	3351 BC	SMU 1151
	Butana (KG23)	4519 ± 67	3283 BC	SMU 1188
	Butana (KG23)	4542 ± 253	3319 BC	SMU 1155
	Butana (KG23)	4727 ± 154	3544 BC	SMU 1201
	Butana (N125)	4410 ± 90	3152 BC	TX 445
Saroba/Kassala Transition	(KG28)	5168 ± 67	4018 BC	SMU1193
Saroba	Malawiya (KG94)	5632 ± 66	4540 BC	SMU 1285
	Malawiya (KG10)	5644 ± 70	4552 BC	SMU 1181
Pre-Saroba	(KG14)	6215 ± 75	5120 BC	SMU 1139
Preceramic	(KG15A)	10,228 ± 273		SMU 1149

Holocene Environments and Occupations in the Southern Atbai

Economic Adaptations

Although the material culture of the various groups within the Atbai Tradition phases tell us relatively little about the articulation between environment, settlement patterns, and economic adaptations, other data are much more instructive. The faunal materials, combined with site location, and site type, are, for the most part, understandable in the context of the locally changing landscape. That is not to say that the changes in the landscape brought about all changes in settlement location or in economic adaptation; rather they provided a changing background into which cultural and economic factors had to be fitted rationally.

Examination of site location by phase (Fig. 5.5), clearly shows a shift in site location and density from the west toward the east: that is, from the area of the Atbara valley and the western steppe to the modern Gash Delta and its western edge. In this regard, there might be a simple correlation between the movement of the Gash to the north-northeast and a movement of site locations to stay close to that river. As will be shown, however, the situation is more complicated.

During the Pre-Saroba Phase, the two known sites consist of small camps situated along either the Atbara River or at the edge of a possible lake, well to the north of the present Gash Delta, near Eriba Station. Economic adaptation was clearly riverine oriented, with an emphasis on, fishing and the exploitation of river edge animals such as hippo, turtle, and crocodile. Animals from wooded savanna regions, such as monkey, warthog, duiker, oribi, and a few of the larger antelopes were also hunted.

A transition to the Saroba Phase is not yet know, since there appears to have been a sudden shift in settlement location from the river edge to the center of the steppe between the Atbara and the Gash (Fig. 5.5). Although it is certain that the Pre-Saroba folk hunted, as well as fished, there is no immediately obvious reason why the river environment should have been abandoned and the focus of occupation shifted to the steppe. In the steppe, the Saroba sites remained small and, if depth of cultural deposits are any indication (Table 5.2), each site was occupied neither long nor intensively. During this phase, the economic base was the hunting of mainly small wooded savanna antelopes, dik-dik, duiker and oribi, as well as warthog, monitor lizard, and an occasional medium-sized bovid, such as kob or reedbuck. Not a single riverine faunal element has been recovered from a Saroba site. In addition, all the Saroba sites have quantities of *Pila wereni* shell present. Unlike the interpretation of their presence at sites near the Nile (Adamson, Clark, and Williams 1974), here it appears that they were not eaten regularly but came to the sites of their own accord during periods of seasonal inundation. In all cases, the piles of shells show many individuals with operculum in place, a sure sign that they were reacting to a drying habitat. Thus, local conditions during the Saroba occupation at least were seasonally very wet and the sites tended to be on slight rises above the otherwise flat steppe. Throughout the Saroba we have no evidence for significant plant exploitation; ground stone is very rare and sickle blades are absent. One must, however, recognize that thin midden deposits at Saroba sites are poorly suited for the preservation of plant remains. It would seem that the site locations of the Saroba Phase peoples were less than optimal. At least some were certainly inundated seasonally but, at best, they appear to have been more or less temporary camps. What seems to have drawn people to this particular area within the Southern Atbai was the rich fauna living in the wet, wooded savanna that would have developed along the banks of the Gash and further from it in the older meander channels it left as it moved to the northeast.

The geographic range of the Saroba peoples is unknown; whether they always inhabited such seasonally wet environments is still to be established. However, they apparently were well adapted because by 4000 BC, their sites increase in size and artifactual density, and new forms of ceramics appear, including large utility vessels and fine ripple-ware beakers. Also, some luxury items such as lip plugs and eggshell beads are found. Site location remains the same but a significant number of grinding stones occur, although there are still no plant remains. In spite of these changes, midden accumulation is only slightly greater (Table 5.2) and there is no evidence for permanent occupation.

Economic adaptation was still based on the hunting of small wooded savanna forms; a few larger animals, including elephants, were taken. Mounds of *Pila wereni* shell attest to the continued seasonal inundation of the central steppe, between the present Atbara and Gash River valleys.

Although there is still a 500-year hiatus in our absolute chronology between the Saroba/Kassala Transitional Phase and the Early Kassala Phase dating to *ca.* 3500 BC, there is no doubt as to the developmental nature of the change. Ceramically, at least, there is marked continuity. Yet radical changes take place in site location, site type, and site size.

Perhaps the most striking change is in settlement patterns; sites are now found both in the Atbara River valley and in the steppe to the east (Fig. 5.5). Also, there are two very distinct site sizes: the majority are rather small, no larger than a hectare (10,000 m^2), while three sites are between 8 and 10 hectares. The difference exceeds size alone (Table 5.2). Midden accumulations at the smaller sites are shallow, the deepest with only *ca.* 35 cm of cultural deposit. The large sites, on the other hand, have cultural materials to a depth of 2 m or more. On the smaller sites,

Holocene Environments and Occupations in the Southern Atbai

ceramics are sparse and lithic materials have only a moderate density. At the larger sites, the density of ceramics is impressive and there is a high frequency of lithic artifacts. The lithic materials include numerous grinding stones, polished stone artifacts (axes, adzes, pointed mace heads, agate lip plugs), all forms which almost never occur at the small sites. Only chipped stone picks occur at both the large sites and the smaller ones near the Atbara. In fact, one of the Atbara sites appears to have specialized in the production of these picks. Thus, not only is there a radical change in site location and site content, but also in site variability from the Saroba to the Early Kassala Phase. This suggests a new social organization, a new way of dealing with the environment, and more complex forms of adaptation. However, on a regional level, no marked change in climatic conditions occurred at this time. The Middle Holocene climate was firmly in place; it was somewhat drier than before but there had been no significant change in either the local flora or fauna.

Table 5.2: Southern Atbai Site Characteristics by Phase or Group

Phase	Number of Sites	Site Areas	Depth of Deposits
Jebel Taka			
Hagiz	27	0–4 ha	0–10 cm
	2	6–12 ha	0–10 cm
Late Kassala			
Jebel Mokram	*ca.* 60	.1–6 ha	0–50 cm
Gash	*ca.* 25	.1–3 ha	0–10 cm
	2	8–11 ha	100–250 cm
Butana	3	.3–2 ha	0–10 cm
	2	4–8 ha	50–150 cm
Early Kassala			
Butana	8	0–4 ha	0–35 cm
	4	9–11 ha	100–260 cm
Saroba/Kassala Transition	1	1.5 ha	*ca.* 30 cm
Saroba	14	.2–1 ha	0–30 cm
Pre-Saroba	2	.1–.3 ha	0–30 cm

The distribution of sites, particularly the larger ones, indicates that location near a single major resource, such as water, was not a concern. The large sites occur on the Atbara, in the drier clay portion of the steppe and, on the eastern edge of their distribution, near the wetter area of the Gash channels, the Shurab el Gash. The size of the largest sites and the depth of their deposits indicate either permanent habitation or regular, repeated occupations which lasted a significant portion of each year.

The smaller sites, with their shallow deposits and the poverty of their material remains, indicate very ephemeral camps which must have been structurally related to the few large sites. There appears to have been two kinds of small sites: small camps, perhaps, related to transitory local exploitation away from the larger sites and those in the Atbara River valley which may have been small villages that specialized in the exploitation of the lithic resources regionally available only in that river valley (chert, agate, greenstone, etc.). An understanding of what the smaller sites signify depends to a large extent upon their dating, since during the Early Kassala Phase, there is an apparent broadening of adaptive strategies.

We suggest that the onset of the Kassala Phase was brought about by an increasingly complex adaptive strategy and that the survival and prosperity of this phase was related to another increase in complexity of the economic base.

The shift to the Kassala Phase, through the transitional period, shows only one significant change in material remains: a marked increase in the number of grinding stones at the Saroba/Kassala Transitional site, as compared with the earlier Saroba sites. However, by the basal Kassala Phase, all the sites are not only rich in ground stone but exhibit the material complexity already described. Within the Central Sudan, the period of the fourth millennium BC is one where animal domestication is well known in the Nile valley (Haaland 1981; Krzyzaniak 1977) and for which many assumptions have been made about sorghum and millet cultivation (Haaland 1981, 1984). Thus, it would not be unreasonable to postulate that the radical changes in site size, site location, and material culture are related to the adoption of domestic plants and animals into the local economy. Yet, the available data preclude such an inclusive interpretation. Instead, it is postulated that only intensive plant exploitation was introduced at that time. Whether or not it involved the introduction of domesticates cannot now be determined. However, it is possible to exclude the introduction of domestic animals as a factor in the change in settlement pattern, site size, or site complexity.

At the major Early Kassala Phase sites, the pattern of faunal recovery was standard; the lower half of each site produced faunal assemblages of wholly wild animals. In fact, these assemblages are not significantly different from those of the Saroba sites which preceded them. However, about halfway up the stratigraphic column, the bones of domestic cattle and small livestock appear in small numbers. At these sites, the radiocarbon dates show little spread (Table 5.1), and it seems as if the appearance of domestic animals took place during the last quarter of the fourth millennium BC. Thus, the pattern of site location, site size, and site complexity was already well established prior to the introduction of domestic animals. It is also important to note that, once introduced, they took quite some time to become a numerically important element in the faunal assemblages.

On the basis of the present evidence, the economic base which supported the Early Kassala Phase florescence in the Southern Atbai initially was one of hunting, with an intensive utilization of plant foods, quite possibly associated with cultivation of cereals. Only in a somewhat later stage did domestic animals play any role. The argument for cultivation can be made on the basis of site location. The large Early Kassala Phase sites are found in the western, or drier portion of the Southern Atbai. Although small numbers of *Pila* were found at these sites, they were individually collected and there is no evidence for seasonal inundation. Thus, by the middle of the fourth millennium BC, the Gash had moved sufficiently to the east to take the central steppe out of the area affected by overbank flooding. This left only the annual summer rains as a source of water. Even today, with only 300 mm of rain yearly, the western steppe is heavily farmed after the summer rains. As one moves eastward into the Sharub el Gash—the area of the rain pools—farming is much less extensive, and tends to be limited to the older high levees of the ancient Gash. It would seem, therefore, that the location of these large Early Kassala Phase sites was partially determined by available arable land. This seems to hold true for the large site along the Atbara as well. It occurs in one of the very few areas where the flood plain becomes wide. So, in this case, it is a matter of land drying sufficiently to provide good cultivation, rather than getting sufficient moisture.

Although the introduction of domestic animals had little initial impact, by the middle of the third millennium BC they had begun to play a major role in the economic base at some sites. Not only are their remains very common, but the absolute number of wild animals hunted drops sharply. This is most noticeable for the antelopes, while sites near the river continued to yield substantial quantities of crocodile, hippo, and warthog bones. These Kassala Phase sites retain their complexity and size and, judging from the number of grinding stones, it would seem that cultivation continued to be a major activity.

By the Late Kassala Phase, perhaps beginning during the end of the third millennium BC, a shift in general settlement location is clearly visible toward the east, in the direction of the present Gash Delta (Fig. 5.5). The major cluster occurs in that area which retains moisture the longest today—the Shurab el Gash and Malawiya rain pools. Most of the large sites (comparable to those of the Early Kassala Phase) have moved even further eastward to along the Gash itself. This suggests that by the end of the third millennium BC the Gash had reached its present bed and the local precipitation had, perhaps, fallen sufficiently to make the western steppe habitable only during and after the summer rains. Site distributions make sense in those terms; small, ephemeral sites with surficial cultural deposits are found over the whole steppe and even into the river valleys, but small to large villages are found only on the eastern half of the steppe and directly around the Gash. Only the paucity of sites within the Atbara valley poses a problem. Here, however, the factor well may have been the limited arable land and a shift to greater differential seasonal flow as the rainfall to the south became more seasonally restricted. The resulting flash floods, a common historic occurrence, would have made the Atbara edge uninviting for permanent occupation.

The Late Kassala Phase is complex for a number of reasons. Unlike earlier periods, there were three distinct cultural manifestations present: one continued the culture seen in the Early Kassala Phase (the Butana Group); one appears to have been an eastern, regional variant thereof (the Gash Group); a third, the Jebel Mokram Group, may have been intrusive into the Southern Atbai but, based on ceramic assemblages, still belonged within the Atbai Tradition. Not only are these groups somewhat different ceramically, their site locations, site types, and site complexity are to some extent different as well (Fattovich, Marks, and Mohammed-Ali 1984).

The continuation of the Early Kassala Phase sites takes place mainly on the eastern fringe of the central steppe, as if there was some reluctance to leave the area. Sites are still large, about 5 to 6 hectares, but only half the size of a millennium earlier. In other ways, however, they retain their identity, although their ceramics become a bit more mundane.

The Gash Group is found mainly at the eastern edge of the Southern Atbai and it is these sites which parallel the Butana Group's massive villages and their rich material culture. These are also like the early Butana Group in the western steppe; there are a few small sites of the Gash Group across the eastern portion of the steppe. Their absence in the western steppe suggests that rather marked territoriality pertained at that time.

The third group, the Jebel Mokram, has the greatest number of sites, the vast majority of which are in the central portion of the steppe, although sites of this group are found from the eastern edge of the Atbara valley to the eastern side of the Gash River (Fig. 5.6). These sites, while they can cover some considerable area (Table 5.2), usually have shallow midden deposits, often consisting of small, discrete concentrations of artifacts. The ceramics fall within the Atbai Tradition but stand out from those of the other groups by virtue of special design motifs and a limited range of surface treatments.

Thus, three identifiably different groups occupied the Southern Atbai at about the same time—from *ca.* 3000 BC until about 1000 BC. While their site distributions overlap to some extent (Fig. 5.6), each tends to have a different settlement pattern. It is suggested that these differences can be understood in terms of historic and economic considerations, as they relate to local conditions.

At the Late Kassala Phase Butana Group sites, domestic animals were known and used but hunting still retained the dominant economic position. While no macrobotanical remains have been recovered, the importance of cultivation is shown by the presence of numerous grinding stones and large numbers of potsherds from thick-walled, large storage vessels.

The fauna from the villages of the Gash Group, situated mainly along the Gash River, contained a small number of domestic animals. However, here, too, hunting seems to have been dominant (Geraads 1983) and fishing common. Floral materials include wild forms such as *Zizyphus* and some *Leguminosae*, as well as some *Hordeum* (Constantini *et al.* 1982). Thus, the economic pattern appears to have been a mixed economy with domestic animals playing a relatively small role and the exploitation of wild plants and animals being most important. In this sense, the economy of the Gash Group is comparable to that of the Late Butana Group, with the addition of fishing and the hunting of riverine animals.

It is within the Jebel Mokram Group that a new economic pattern can be documented. Although faunal preservation is poor, owing to the shallow depth of midden deposits, all the bones so far recovered can be attributed to domestic cattle. In addition, impressions of both domestic sorghum and millet have been found on thick undecorated sherds and actual grains of domestic millet have been recovered from some sherds (Constantini, personal communication). Thus, the Jebel Mokram

Group would appear to have a mixed cattle herding, farming economy, very much like that of some of the present-day inhabitants of the steppe of the Southern Atbai (Barbour 1964). Only the presence of two possible villages on the eastern edge of the area suggests a somewhat more stable residency for the Jebel Mokram Group than for later peoples.

Since so few of the sites have been dated by radiocarbon, seven out of over 200, it is difficult to fully appreciate the fate of the three groups of the Late Kassala Phase. The general trend suggests that village life was giving way to semi-nomadic patterns as domestic animals became more important and as the central steppe became drier. It seems clear, however, that along the eastern edge of the Southern Atbai, adjacent to the Gash, village life remained viable.

There is something of a gap between the Late Kassala Phase and the Jebel Taka Phase which follows it. The latter is not yet securely dated but on the basis of some ceramic traits it has been placed between 500 BC and 300 AD (Fattovich in press b).

The distribution of Jebel Taka Phase sites is as geographically wide as for any group. The highest density lies near the Malawiya rain pools, an area which provides water well into the dry season (Fig. 5.5). Most sites are small but even the larger ones consist of little more than an even, sparse scatter of cultural materials. No economic data are available but the settlement pattern, the site conditions, and the paucity of artifacts at the sites all suggest that economic adaptation centered mainly on herding with a small amount of cultivation on the steppe after the rainy season. The dry season would probably have been spent near the Atbara and Gash Rivers, accounting for the somewhat more substantial sites in those areas.

It seems that the Jebel Taka Phase is followed by a hiatus in occupation which lasts until after the first millennium AD. While it might be nice to correlate this with increasing dessication, more likely it relates to a well-documented punitive expedition into the Southern Atbai by the Axumite King Exana I in the fourth century AD (Kobishchanov 1979:65).

Summary and Conclusions

Reconstructions of regional paleoenvironments for the Central Sudan show a gradual drying trend since the Early Holocene. The reconstruction of culture change and continuity in the Southern Atbai, on the other hand, shows no such gradualism. Instead, there are periods of rapid change especially in settlement systems and adaptive strategies. On present evidence, these changes can be related to two factors: relatively drastic changes in the local environment of the

Southern Atbai brought about by changes in the riverine hydrography and, secondly, by additions to the economic base of the inhabitants of the Southern Atbai, including the introduction of domesticated plants and animals probably from outside the area.

Four major periods of change can be isolated in the Holocene cultural sequence of the Southern Atbai. First, a change in settlement patterns from the riverine-oriented Pre-Saroba Phase to a wooded savanna-oriented Saroba Phase, all within a hunting/gathering economic base. Second, a radical change in settlement organization, and presumably social complexity, between the Saroba Phase and the Early Kassala Phase, possibly related to additions to the economic base, including a shift to the intensive cultivation of domesticated plants. Third, a diversification in cultural expressions and adaptive strategies, accompanied by a major population increase in the Late Kassala Phase. This change is related, at least partially, to an increase in herding of domesticated animals. Finally, a change from the agro-pastoralist adaptations in the Late Kassala Phase, to a mainly pastoralist and possibly nomadic, adaptation in the Jebel Taka Phase, accompanied by a decline in cultural variability and population levels. Overall, even though the changes in the local hydrography elicit clearly visible changes in general settlement patterns (a shifting site density from the western to the eastern steppe), only a few of the instances of major cultural change can be directly linked to these environmental changes.

Acknowledgements

The field work in the central and western parts of the Southern Atbai has been funded by two National Science Foundation Grants to the Senior author with additional support from the University of Khartoum. Field work in that area was under the direction of the Senior author and Dr. Abbas Mohammed-Ali, Department of Archaeology, University of Khartoum. Faunal identifications, unless otherwise referenced, were made by Mr. Juris Peters, Geologisch Instituut, University of Gent.

References

Adams, W. Y. 1977. *Nubia: Corridor to Africa.* Aldine Press.

Adamson, D., Clark, J. D. and Williams, M. A. J. 1974. Barbed bone points from Central Sudan and the age of the Early Khartoum tradition. *Nature* 249:120–23.

Arkell, A. J. 1949. *Early Khartoum.* London: Oxford University Press.

Barbour, K. M. 1964. *The Republic of the Sudan: A Regional Geography*. London: University of London Press.

Coltorti, M., D'Alessandro, A., Fattovich, R., Lenoble, P. and Sadr, K. 1984. Gash Delta Archaeological Project, 1984 field season. *Nyame Akuma* 24/25:20–3.

Constantini, L., Fattovich, R., Pardini, E. and Piperno, M. 1982. Preliminary report of archaeological investigations at the site of Mahal Teglinos (Kassala), November, 1981. *Nyame Akuma* 21:30-3.

_____., Fattovich, R., Piperno, M. and Sadr, K. 1983. Gash Delta Archaeological Project, 1982 field season. *Nyame Akuma* 23:17–19.

Durante, S., Fattovich, R. and Piperno, M. 1980. Archaeological survey of the Gash Delta, Kassala Province. *Nyame Akuma* 17:64–71.

Fattovich, R. In press a. The late prehistory of the Gash Delta, Sudan. 2nd International symposium on the late prehistory of the Nile Basin and the Sahara, Poznan, 1984.

_____. In press b. The Gash Delta between 1,000 B.C. and 1,000 A.D. 5th International Conference for Meroitic Studies, Rome, 1984.

_____. 1982. The problem of Sudanese-Ethiopian contacts in antiquity: status questions and current trends of research. In *Nubian Studies* (ed J. Plumley): pp. 76–86.

Fattovich, R. and Marks, A. E., and Mohammed-Ali, A. 1984. The archaeology of the eastern Sahel, Sudan: Preliminary results. *The African Archaeological Review* 2:173–88.

Fattovich, R. and Piperno, M. 1981. Survey of the Gash Delta (November 1980). *Nyame Akuma* 19:26–31.

Geraads, D. 1983. Faunal remains from archaeological sites in the Gash Delta, Sudan. *Nyame Akuma* 23:22.

Grove, A. T. 1973. Desertification in the African environment. In *Reports on the 1973 Symposium, Drought in Africa* (D. Dalby and R. J. Harrison-Church, eds.). University of London, School of Oriental and African Studies.

Grove, A. T. and Goudie, A. S. 1971. Late Quaternary lake levels in the Rift valley of southern Ethiopia and elsewhere in tropical Africa. *Nature* 234:403–5.

Haaland, R 1981. *Migratory herdsmen and cultivating women. The structure of Neolithic seasonal adaptation in the Khartoum Mile environment.* Bergen.

———. 1984. Continuity and discontinuity. How to account for a two-thousand-year gap in the cultural history of the Khartoum Nile environment. *Norwegian Archaeological Review* 17(1):39–52.

Kobishchanov, Y. M. 1979. *Axum.* Pennsylvania State University Press.

Krzyzaniak, L. 1977. New light on early food production in the central Sudan. *Journal of African History* 19(2):159–72.

Marks, A. E. 1984. Butana Archaeological Project, 1983–84. *Nyame Akuma* 24/25:32–33.

Marks, A. E. and Fattovich, R. In press. The later prehistory of the Eastern Sudan: A preliminary view. International symposium on the late prehistory of the Nile Basin and the Sahara, Poznan, 1984.

Mohammed-Ali, A., and Marks, A. E. 1984. The prehistory of Shaqadud in the western Butana, central Sudan: A preliminary report. *Norwegian Archaeological Review* 17(1):52–62).

Sadr, K. 1984. The Gergat Group: The latest archaeological phase in the southern Atbai, east central Sudan. *Nyame Akuma* 24/25:33–5.

Warren, A. 1970. Dune trends and their implications in the central Sudan. *Zeit. Geomorph., Suppl.* 10:154–80.

Wickens, G. E. 1975. Changes in the climate and vegetation of the Sudan since 20,000 B.P. *Boissiera* 24:43–65.

Wickens, G. E. 1982. Paleobotanical speculations and Quaternary environments in the Sudan. In *A Land between Two Niles: Quaternary Geology and Biology of the Central Sudan* (eds M. A. J. Williams and D. A. Adamson): pp. 23–51. Rotterdam: A. A. Balkema.

Williams, M. A. J. and Adamson, D. A. 1974. Late Pleistocene dessication along the White Nile. *Nature* 248:584–585.

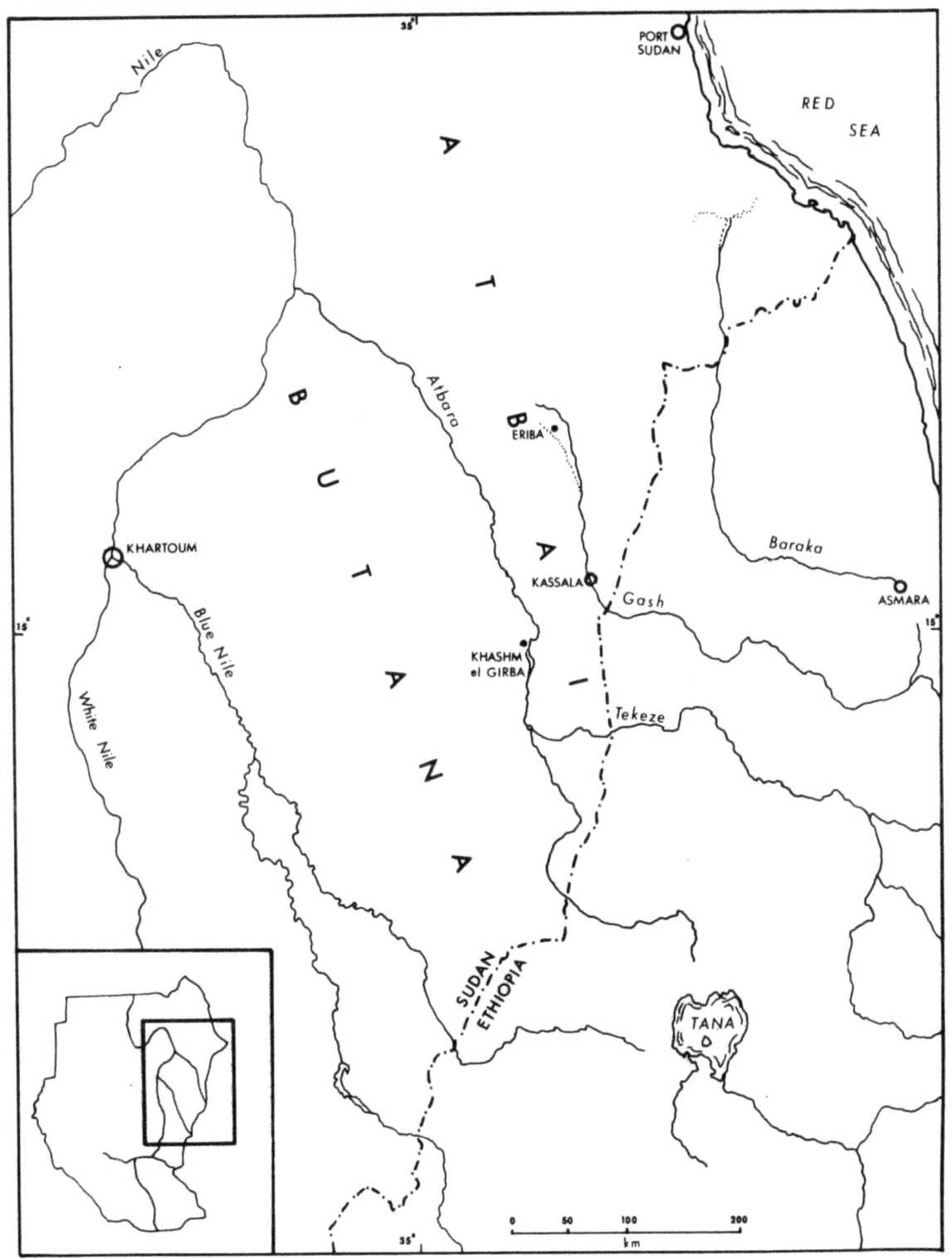

Fig. 5.1: Map of the Eastern Sudan.

Holocene Environments and Occupations in the Southern Atbai

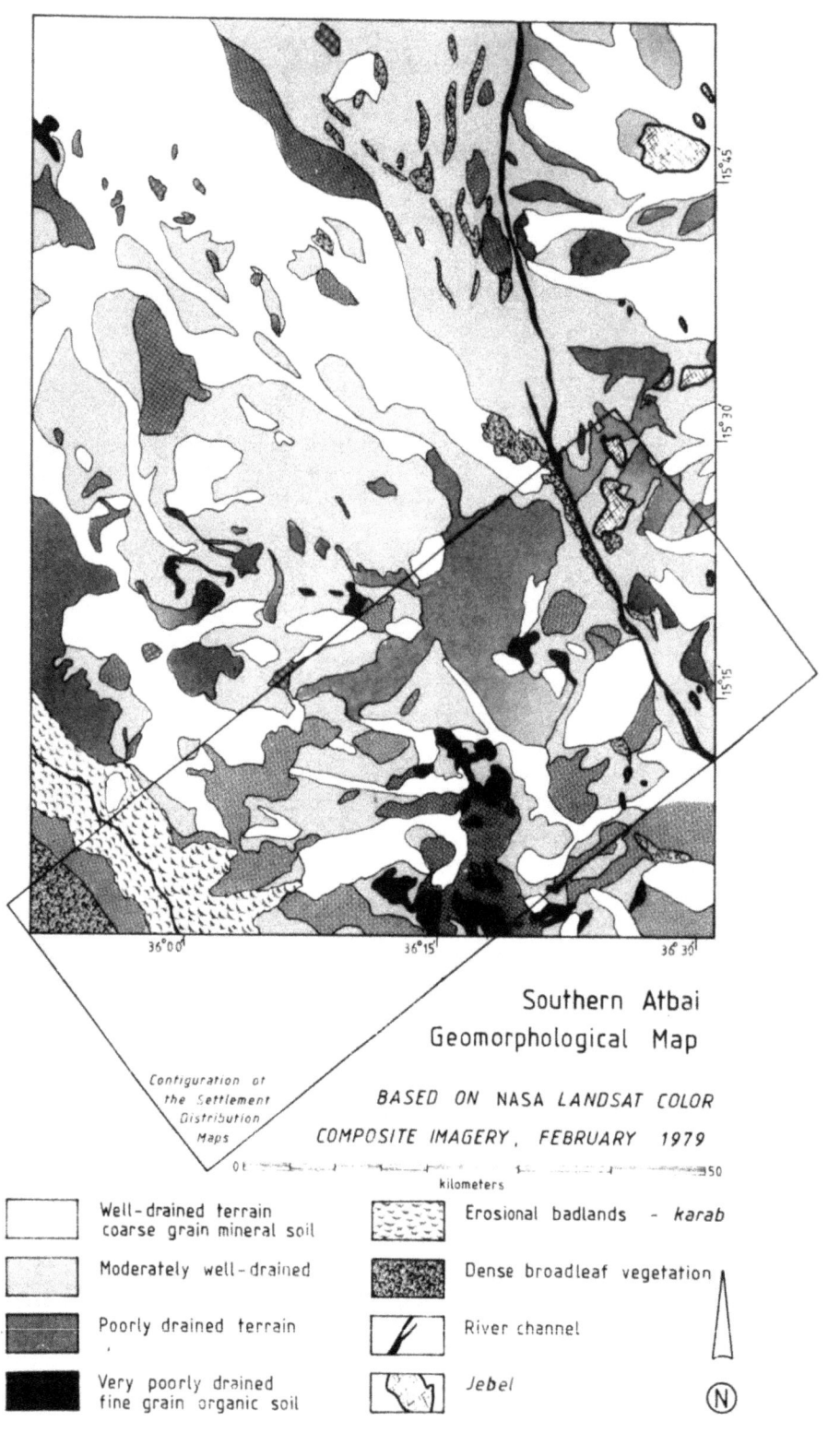

Fig. 5.2: Geomorphological Map of the Southern Atbai; simplied from NASA Landsat imagery; with outline of survey zone superimposed.

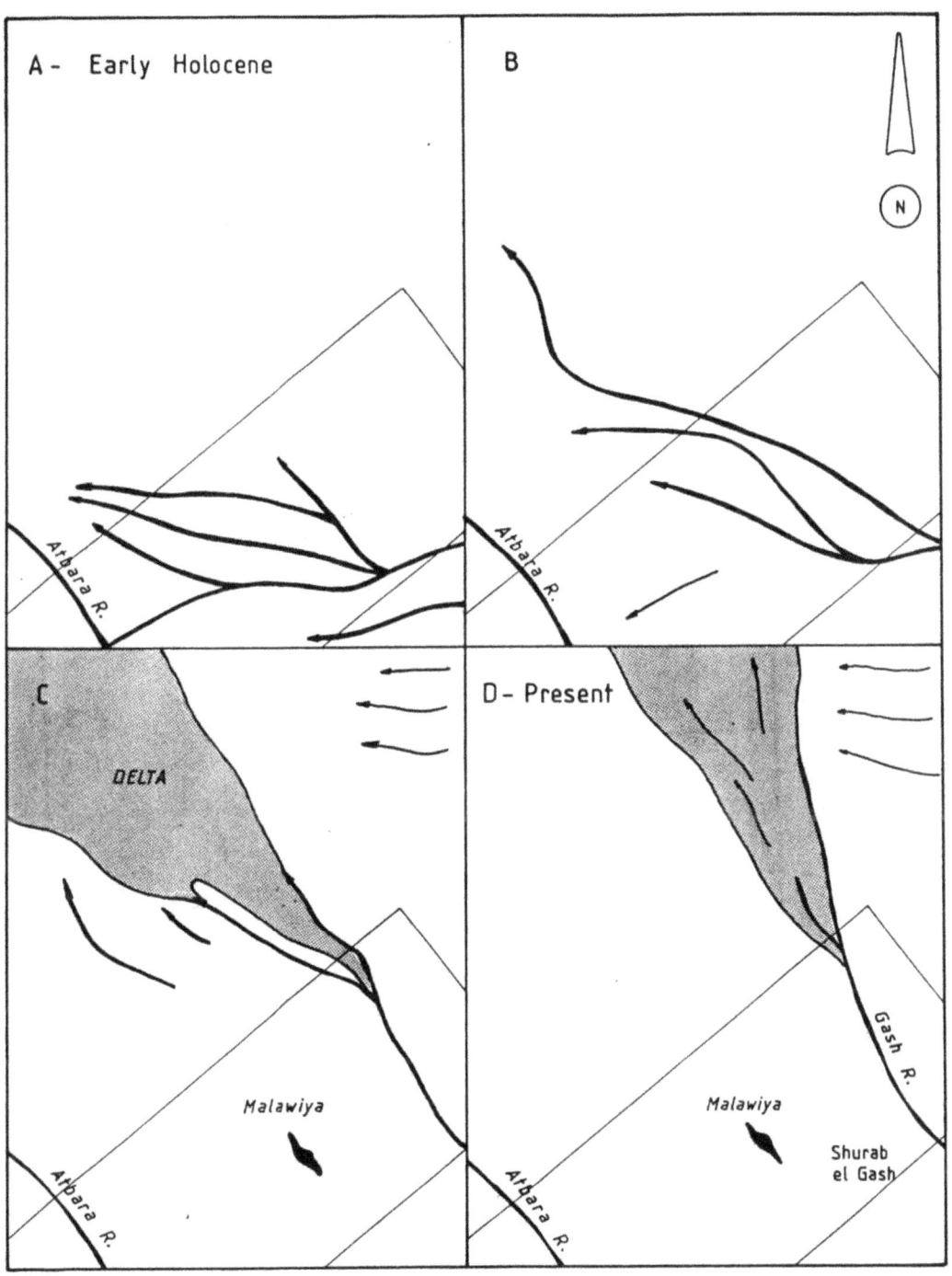

Fig. 5.3: Stages in the hydrography of the Gash River; hypothetical reconstruction based partially on information from NASA Landsat imagery. The dates for stages B and C are unknown, but fall within the Middle Holocene.

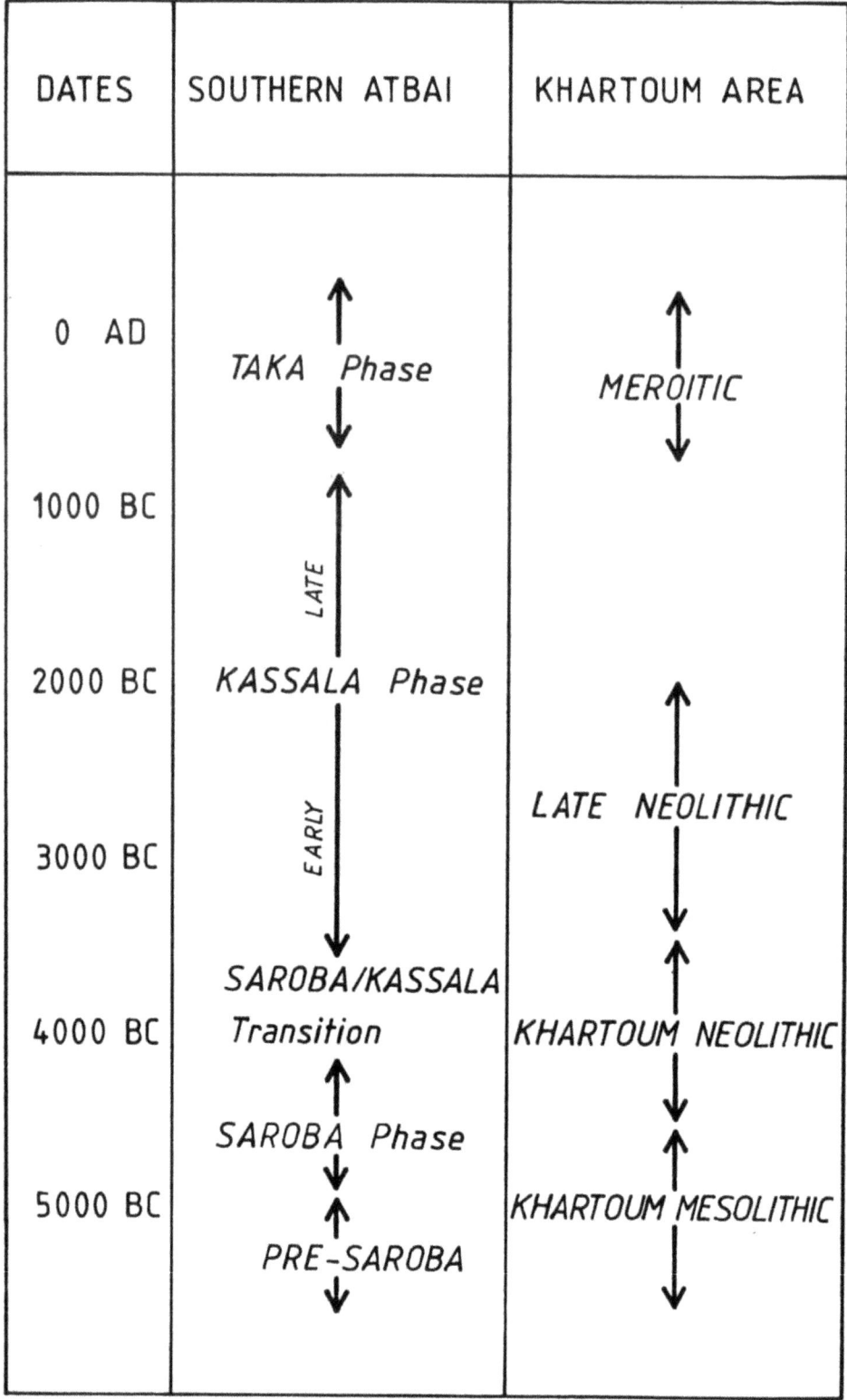

Fig. 5.4: Correlation of the Southern Atbai sequence with that of the Central Sudan in the Khartoum area.

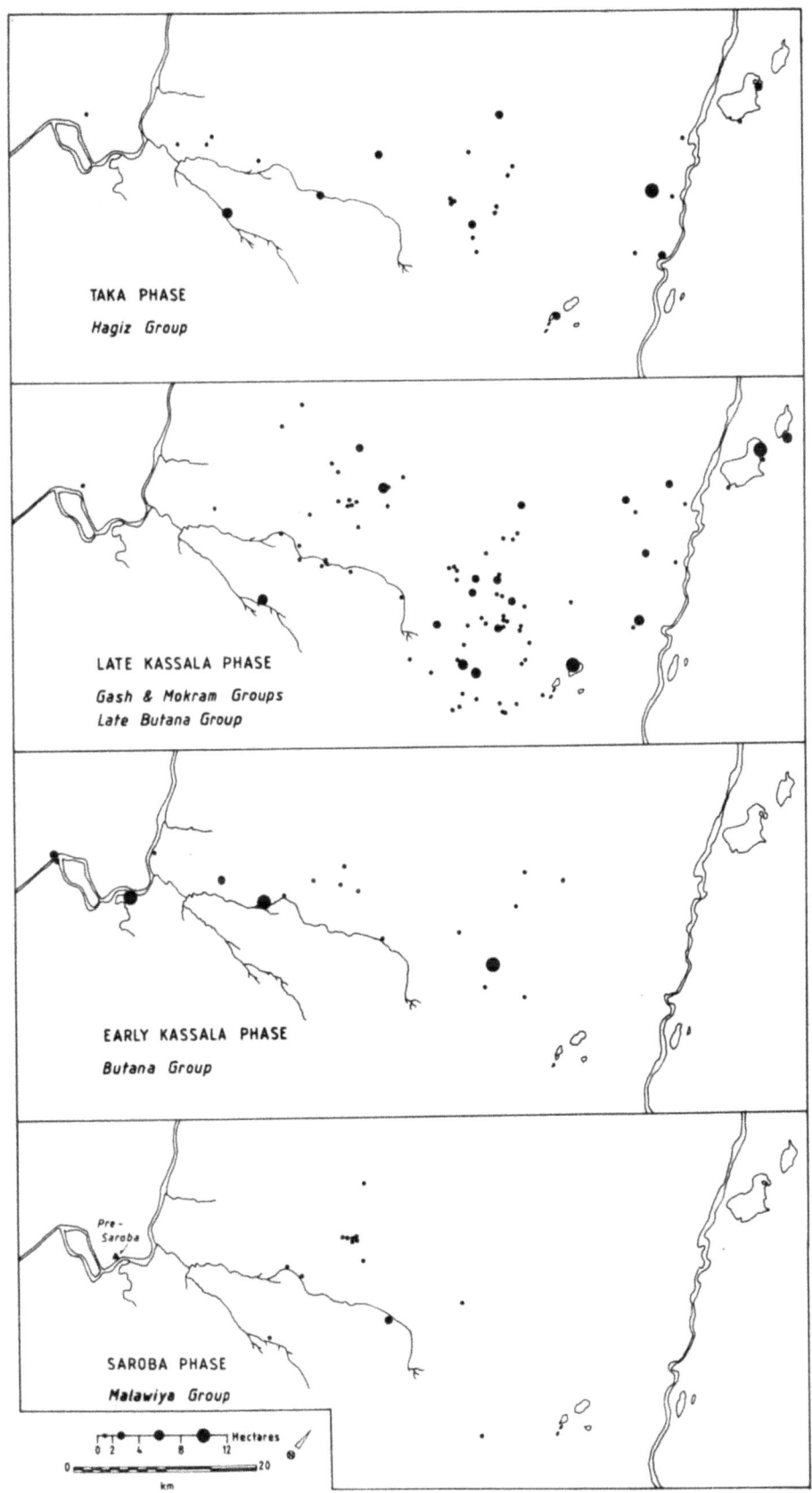

Fig. 5.5: Site distribution by phase, within the survey zone of the Southern Atbai.

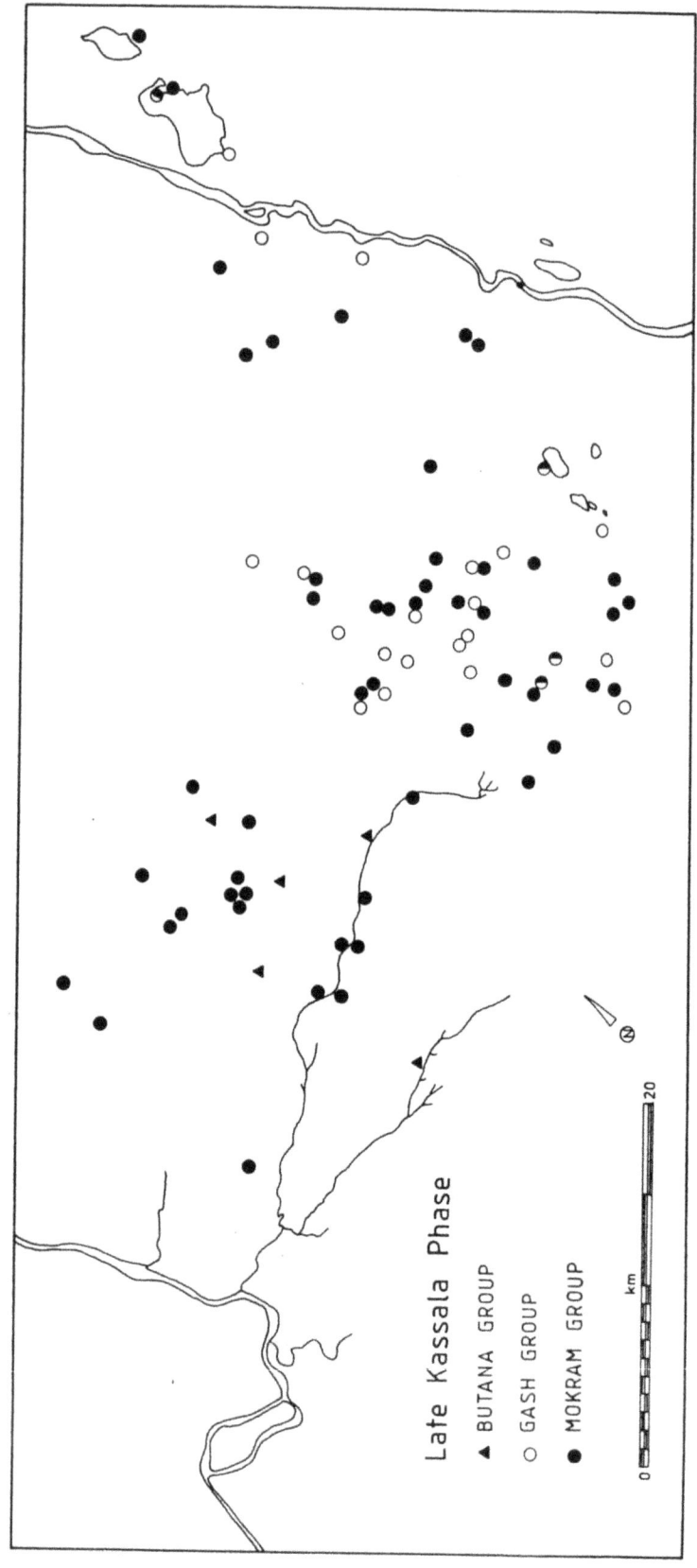

Fig. 5.6: Distribution of sites during the Late Kassala Phase by Group.

Evolution of Stone Age Food-Producing Cultures in East Africa

By John Bower, Department of Sociology and Anthropology, Iowa State University, Ames, Iowa.

"The most that can be said of major changes in subsistence practices, economic activity and socio-political organization is that, once they are triggered, their general course may depend on endogenous conditions as much as, or more than, it does on the nature of some outside stimulus." (Tolstoy 1987)

This chapter deals with Holocene cultures in Kenya, Tanzania and Uganda which, though possessing domesticates, lacked metallurgy. Such cultures have been assigned to the Pastoral Neolithic (PN), a regional developmental stage that is particularly relevant to the theme of this volume since it encompasses major transformations in both human life-ways and the habitats where they were practiced. The paper begins with a review of the lore regarding the nature and history of the PN, continues with a discussion of new evidence concerning PN taxonomy, chronology and geography and concludes with a reinterpretation of the evolutionary significance of the PN based, in part, on comparisons with the histories of broadly similar stages in other parts of Africa.

Pastoral Neolithic Lore

The current state of opinion regarding PN taxonomy, chronology and geography (by which I mean not only the spatial distribution of PN entities but also their subsistence patterns, settlement systems, population densities, etc.—in a word, their ecology) is to say the least, unsettled (Phillipson 1982). This is partly because data concerning the morphology and spatiotemporal distribution of PN artifacts, as well as evidence of PN subsistence-settlement behavior, are exceedingly spotty (cf. Phillipson 1977; Ambrose 1984a.) It is also partly a consequence of two recent, well-intentioned but essentially premature attempts to systematize the PN, the outcomes of which are largely at odds with each other, thus intensifying the confusion that surrounds PN studies. In this section, I shall briefly survey the development of contemporary PN lore with a view toward identifying areas of broad agreement, as well as strongly divergent views. The discussion will focus on two basic issues: the culture-historical framework for the PN and the major features of the evolution of PN cultures.

The existence of a "neolithic" stage in East African prehistory was recognized by Louis Leakey as early as the 1930s (L. S. B. Leakey 1931); but the establishment of a comprehensive identity for the stage had to await the results of Mary Leakey's investigations at Hyrax Hill (M. D. Leakey 1945), where numerous stone bowls were recovered. The occurrence of such bowls in various assemblages with flaked tools of Later Stone Age (LSA) type and diverse ceramic traditions led to the recognition of the "Stone Bowl Cultures" as a formal "neolithic" entity (Cole 1963). Thus, until about the mid 1970s, "neolithic" research in East Africa centered largely on determining the nature of the "Stone Bowl Cultures" (cf. Cohen 1970; Odner 1972; Bower 1973).

However, as "neolithic" data accumulated, it became apparent that some occurrences which matched one or more "Stone Bowl" sites in pottery and/or flaked stone typology lacked the culture's *fossile directeur*, a form closely identified with mortuary ritual and, perhaps for this reason, relatively uncommon in habitation debris. At the same time that the taxonomic significance of stone bowls was declining, attention was being increasingly directed toward the fauna from the "Stone Bowl" sites, particularly the domestic livestock (Gramly 1975; Onyango-Abuje 1977). Among the more compelling developments behind this shift in perspective were the recognition that the domestic fauna (cattle, sheep and goats) represented the earliest occurrence of food production in East Africa, that the cultures which raised them could therefore be regarded as "neolithic" in the economic sense of the word and that the East African "Neolithic" might, in some measure, constitute an autochthonous development worthy of comparison with other, more or less independent "neolithics" in different parts of Africa and even in other continents.

This trend was given formal recognition in the late 1970s by the abandonment of the "Stone Bowl Cultures" as an entity and the establishment of a developmental stage called the Pastoral Neolithic (Bower *et al.* 1977:119). Since the meaning of the term Pastoral Neolithic seems to vary depending upon who is using it, let me quote its original definition: "Pastoral 'Neolithic' (PN) is used to refer to societies with an LSA technology and a pastoral economic base relying heavily on domestic cattle and/or ovicaprids." The discussion that accompanied the definition (Bower *et al.* 1977:119) made it clear that the PN was to be regarded as an evolutionary stage intermediate between the LSA, lacking domesticates, and the Pastoral Iron Age (PIA), in which a reliance upon domestic livestock co-occurs with metallurgy. Two points deserve special emphasis: 1) the defining characteristic of the PN is the combination of domesticated fauna and stone age technology, so that the presence or absence of stone bowls, pottery, stone axes, etc. is incidental; 2) the PN is conceptually parallel to LSA, MSA, etc. and does not refer to a cultural entity, such as an industry (cf. Bishop and Clark 1967:879–901; Clark *et al.* 1966).

As a result of the intensification of PN field studies during the 1970s, it was apparent by the latter part of the decade that the stage contained several, perhaps many, markedly different ceramic styles with varied, though often poorly defined, spatiotemporal distributions, some of which overlapped on either the spatial or the temporal dimension but usually not both. Since these ceramic variants (or wares to use the colloquial term) seemed to point toward the existence of distinct cultural entities within the PN stage, the wares were formally defined by Wandibba (1977; see also Bower *et al.* 1977) with the idea that they might serve as a foundation for a PN taxonomy. A total of six wares was recognized: Nderit, Narosura, Remnant, Akira, Maringishu and Kansyore (see Table 6.1). Each of these is not only typologically distinct but (with the exception of Kansyore) also occupies a reasonably discrete position in space and/or time, as will be shown later. Thus, on both typological and distributional grounds, the wares seemed to provide a sound basis for delineating PN entities.

Table 6.1: Cultural Entities in the East African PN

Wandibba	Collett/Robertshaw	Ambrose
Akira		SPN/Eburran
Maringishu		SPN/Eburran
Remnant (Elmenteitan)	Elmenteitan	Elmenteitan
Narosura	Oldishi	SPN/Eburran
Kansyore	Oltome	—
Nderit	Olmalenge	SPN/Eburran

Note: Entities listed under Wandibba are in chronological order, from oldest-youngest reading up, although Narosura, Elmenteitan and Maringishu wares may be penecontemporaneous; see text for a discussion of the chronology. Entities under Collett/Robertshaw and Ambrose are aligned with their correlates in Wandibba's classification.

In summary, by the end of the 1970s, the culture-historical framework for the study of stone-age food production in East Africa consisted of a developmental stage, called the Pastoral Neolithic, that contained six more or less well defined cultural entities (ceramic wares) and, given the spottiness of PN data, admitted the possibility of discovering additional taxa through future investigations. However, the validity of this framework was seriously challenged in the early 1980s (Ambrose 1984a, b; Collett and Robertshaw 1983; Robertshaw and Collett 1983a). Though the attacks from different sources yielded different results, their initial targets for criticism were the same:

1) the validity of the PN as an entity and

2) the credibility of its alleged subdivisions (ceramic wares).

Evolution of Stone Age Food-Producing Cultures in East Africa

The first criticism is the easiest to deal with because it stems from a basic conceptual misunderstanding of the PN. As I have shown earlier, the PN was defined as a regional developmental stage, intermediate between the LSA and the PIA. Since it was obviously never intended to be treated as a cultural entity, it should not be criticized for what it was not supposed to be. One can, however, ask whether the PN's status as a developmental stage is warranted; this question was not considered in the cited critiques, but will presently be addressed.

Given that the PN is not a cultural entity, the attack on its presumed "subdivisions" is also, to some extent, spurious. However, since the criticism was ultimately concerned with the entities defined by Wandibba (1977) *qua* entities, I will respond to it on this level. That is, I will discuss the critique of Wandibba's PN wares and the alternative taxonomies proposed by Ambrose and Collett/Robertshaw.

The views of the latter were published earlier, so I will begin with them. Among the more significant of David Collet and Peter Robertshaw's comments on PN taxonomy are those which refer to the nature of the cultural entities and those which relate to the data used in constructing them. As regards the nature of the entities, while Collett and Robertshaw (1983) agree that variation in ceramic "style," which they define as standing "in contrast to function," is an appropriate basis for constructing cultural entities, they object to linking such entities to "wares," which they claim are "something midway between an assemblage and a type." As regards the data used in constructing entities, these authors point to problems of sample size and chronological control. In particular, they note that many PN assemblages contain meager collections of pottery which cannot be dated relative to one another (i.e., stratigraphically) and are inadequately dated on chronometric grounds.

The remedy proposed by Collett/Robertshaw (1983) for dealing with the perceived deficiencies in the PN taxonomy consisted of erecting units on the basis of an average-link cluster analysis of selected attributes (decorative motif and technique, vessel form, etc.) on a total of 708 reconstructible vessels from 16 sites. The units so identified were claimed to be conceptually comparable with the kind of entity that Willey and Phillips (1958) describe as a tradition, and four such units were named: Elmenteitan, Olmalenge, Oltome and Oldishi. All but one of the "tradition" names (Elmenteitan) are "nonsense names with an African flavour" (Collett and Robertshaw 1983:121).

It is noteworthy that, with two exceptions, the "traditions" erected by Collett/Robertshaw correspond with Wandibba's wares (Table 6.1). One of the exceptions, Wandibba's Akira ware, was unrepresented in the

Collett/Robertshaw analysis, which included no sherds from any of the Akira occurrences. The other exception, Wandibba's Maringishu ware, is subsumed within Collett/Robertshaw's Olmalenge "tradition." I believe this is a consequence of their having treated the Hyrax Hill assemblage (M. D. Leakey 1945) as a single component occurrence, whereas the collection clearly contains at least two markedly different ceramic styles (cf. Ambrose 1984b:142). One of these is obviously related to Maringishu ware, exhibiting a highly distinctive Maringishu decorative motif (Bower et al. 1977:Fig. 2, f and g), and another is closely comparable with Nderit ware. Thus, the Olmalenge "tradition" is a result of conflating distinct ceramic wares.

In general, I agree with Collett/Robertshaw's comments regarding deficiencies in the data upon which Wandibba's wares are based and I also share, in part, their views about conceptual problems in the application of the term ware to PN ceramics. However, I fail to see how any of these difficulties can be overcome by applying an elaborate statistical analysis to a very small aggregate of pottery in order to establish a new taxonomy that is conceptually derived from New World archaeology. It seems to me that the most useful and convincing outcome of the Collett/Robertshaw study is its general confirmation (with the exceptions noted earlier) of the entities recognized by Wandibba. Finally, while I agree the ware originally called Remnant (which refers to a person) should be renamed Elmenteitan (which refers to a place), I see no advantage in substituting "nonsense" labels for entity names (e.g., Nderit, Narosura, etc.) that are not only deeply entrenched in the PN literature but also refer to type localities. In fact, such substitution can only add to the confusion that is widely recognized as the bane of PN studies.

I turn next to the views of Stanley Ambrose whose syntheses of stone age food production in East Africa (Ambrose 1984a, b) are by far the most exhaustive and carefully developed statements on the subject so far published. On taxonomic issues, Ambrose takes the position that entities should be defined on the basis of variation in the characteristics of flaked stone tool assemblages. (He claims to employ a polythetic approach to taxonomy, but his primary source of definition is clearly lithic material. Indeed, so strong is his commitment to the "lithic standard" that he is forced to explain away contradictory evidence from ceramic materials, as will presently be shown.) Chiefly on the basis of differences in manufacturing technique and morphometric attributes of formal tools, Ambrose has created three taxa (or industries) relevant to the present discussion: the Eburran (which has its origins in the LSA, but overlaps the PN stage), the Savanna Pastoral Neolithic (SPN) and the Elmenteitan.

Evolution of Stone Age Food-Producing Cultures in East Africa

There are two major problems with this scheme, one theoretical and the other empirical. As regards the former, the cultural significance of lithic taxonomies is undermined by the as yet unresolved theoretical problem of recognizing "style" in lithic artifacts (Sackett 1982). By comparison, the cultural significance of ceramic variation is relatively straightforward.

The most damaging empirical problem with the Ambrose scheme is that, as Table 6.1 shows, each of his taxa embraces several markedly different ceramic wares, the only exception being the Elmenteitan industry. Ambrose, himself, is at least indirectly vexed by this problem when he finds inordinate ceramic variety at a site he attributes to a single lithic industry. Thus, for example: "The diversity of pottery types at Hyrax Hill is intriguing, and suggests to me that Eburran populations, in the process of adopting a pastoral lifestyle, interacted with a variety of Savanna Pastoral Neolithic groups, perhaps acting as middlemen and entrepreneurs, in order to gain a foothold in the pastoral economy" (Ambrose 1984b:316). Not only does this statement imply recognition of the cultural significance of ceramic variation, it also strikes me as a rather labored interpretation of what, on cultural, stratigraphic and chronological grounds (cf. Ambrose 1984b:143), appears to be a multiple component site with low rates of sedimentation.

If I seem to have devoted a great deal of space to PN taxonomic issues, it is because this is the domain of inquiry about which there is the greatest amount of disagreement. Turning to questions regarding the origin and spread of the PN, one finds a reasonably strong consensus on most points, with an occasional sharply divergent minority opinion. For example, most workers seem to agree that the PN originated in northern portions of East Africa at about 4000 bp, presumably as a result of diffusion from either the Sudan or southwestern Ethiopia (cf. Phillipson 1977, 1982). However, a few dissenters argue for a much earlier origin, perhaps in excess of 7000 bp (cf. Bower and Nelson 1978). It also appears to be broadly accepted that, whatever the ultimate source of PN livestock might have been, the process whereby they were introduced to East Africa involved limited human immigration, since there is substantial evidence of cultural continuity between the regional LSA and PN stages and little evidence of cultural ties with areas to the north.

The prevailing view of the spread of the PN, though generally not explicitly formulated, seems to entail a southward flowing "bow wave" model (cf. Williams, chapter 3) in which the "wave's" rate of travel is at least partially determined by the mid-late Holocene shrinkage of the Gregory Rift lakes (Butzer *et al.* 1972; Richardson and Richardson 1972), exposing vast areas of previously inundated grazing land. Such an ecological perspective is also evident in Ambrose's views regarding the

Eburran, SPN and Elmenteitan industries, each of which is shown to have a somewhat different distribution with respect to elevation (Ambrose 1984a:Fig. 5) and suggested to represent adaptations to different types of habitat (1984a:228-9). While this may be true of Ambrose's lithic "industries," thus underscoring the possibility that they may be at least partly shaped by function rather than style, Wandibba's ceramic wares exhibit no such geographic correlation, occurring instead in a remarkably wide range of ecozones (Bower 1978; 1984a, b).

A Fresh Perspective on Pastoral Neolithic Culture History

In this section, I shall offer a revised view of Pastoral Neolithic culture history and geography, taking into account both existing lore and new data from recent work in the Serengeti plains, Tanzania, and elsewhere. Turning first to the PN's status as a developmental stage intermediate between the Later Stone Age and the Pastoral Iron Age, it seems to me that this remains an essentially sound and useful evolutionary construct. However, in light of recent evidence of marked variation in the incidence of domestic fauna at PN sites (Gifford-Gonzalez and Kimengich 1984) and the possibility that domesticated grain, as well as livestock, were used, I propose modifying the definition as follows: "Pastoral Neolithic refers to societies with an LSA technology, domestic livestock, and a wide variety of subsistence regimes." Further modification may be necessary in future, but this definition provides common ground for discussing an important segment of East Africa's culture history, which is about the best one can expect of any such construct.

In a similar vein, I would argue that, despite the shortcomings of the data from which Wandibba's ceramic wares are derived, the wares offer the best available framework for defining PN cultural entities. Not only do they stand up to independent statistical testing, as shown by the outcome of the Collett/Robertshaw analysis discussed earlier, but they also appear to have essentially non-overlapping boundaries in space and/or time. It seems clear that additional wares will be recognized in future; at least two, Salasun and Ileret wares, have already appeared in the literature (Ambrose 1984b; Barthelme 1985) and recent finds in the Serengeti (Bower and Chadderdon 1986; Bower and Smale n.d.) may yield others. So far, however, new data have required elaboration, but not replacement of Wandibba's scheme.

The relationship between Wandibba's wares and Ambrose's lithic industries is, in my opinion, problematic. Although Ambrose (1984a, b; see also Table 6.1) seems to believe that he has worked this out, my own as yet incomplete investigations point toward conclusions that are considerably at odds with his. For example, I am inclined to see a close link between Narosura (possibly also Akira) ware and the Elmenteitan industry (cf. Mehlman 1977:113–115), though I agree that the Elmenteitan industry, alone, is associated with Elmenteitan ware. I also feel that Maringishu and Nderit wares may share a common industry quite different from that associated with any of the other PN wares. However, since the present state of my research does not permit me to demonstrate such relationships, I will simply register my reservations concerning Ambrose's results and admit that I have not yet got anything defensible to offer in their place.

Contrary to Robertshaw and Collett's (1983a:293) pessimistic outlook on PN chronology, there are numerous reasonably useful lines of evidence from which both relative and absolute temporal information can be derived. Thus, for example, stratigraphic relationships at multiple component sites on Lukenya Hill (GvJm44; Bower *et al.* 1977), at the Seronera Game Lodge (Bower 1973) and at the Gol Kopjes (Bower and Chadderdon 1986; Bower and Smale n.d.), suggest that Nderit ware is older than Kansyore, Narosura and Akira wares. Moreover, Robertshaw and Collett's tendency to reject summarily any radiocarbon determination based on bone apatite while insisting that charcoal dates provide the only suitable basis for chronometry ignores both the convergence of evidence in support of apatite dates and their own caveat concerning charcoal dates (1983a:294): "There remains always the question of the validity of the association between the charcoal sample and the archaeological material to which the date is believed to refer." This cautionary note is particularly germane to comminuted charcoal samples from poorly consolidated sediments such as frequently occur at PN sites. In view of such problems, I believe it is necessary to evaluate all chronometric information carefully, rather than simply to reject bone apatite dates and accept those based on charcoal.

The temporal position of several PN wares has been reasonably well established, while others are more tenuously dated. Thus, Narosura ware seems to occupy most of the third millennium bp (Odner 1972), while Elmenteitan ware is confined to the last half of the same millennium (Bower *et al.* 1977; Ambrose 1984a) and Akira ware apparently spans the interval between about 2000 and 1300 bp. (Bower *et al.* 1977). Maringishu ware is represented by but one radiocarbon date of 1695 ±105 bp (GX-4466A) that suggests temporal overlap with Akira ware. Chronometric data for Kansyore ware are both sparse and insecure (Collett and Robertshaw 1980; Robertshaw *et al.* 1983); however, available estimates suggest a time span from about 5000 to 2600 bp (Mehlman 1977, 1979; Soper and Golden 1969). Finally Nderit ware has associated radiocarbon dates ranging from about 4000 bp at the northern limit of its known geographic distribution (Barthelme 1984) to about 7000 bp at its southernmost occurrence in the Serengeti Plains. Although the latter date is based on bone apatite determinations, several lines of evidence tend to support it (Bower and Chadderdon 1986; Bower and Smale n.d.). Thus, I believe the date should be tentatively accepted, pending additional chronometric information on Nderit ware.

The spatial boundaries of the PN wares are perhaps rather less well defined than their temporal limits, but it is at least possible to describe the geographic distribution of most of them in general terms. Nderit ware occupies a vast expanse of land in Kenya and Tanzania, with a north-south axis more or less centered on the Gregory Rift, reaching at least as far north as Lake Turkana and as far south as the Serengeti Plain. On the east, the entity stretches at least as far as Lukenya Hill, about 50 km east of Nairobi, and, on the west, at least as far as the northwest corner of the Serengeti National Park, in the Lake Nyanza (formerly Victoria) basin. With one exception, this is by far the most extensive PN area in East Africa; only Akira ware is distributed over a more or less comparable amount of land.

Most known occurrences of Kansyore ware are limited to the Lake Nyanza basin, both east and west of the lake. However, the ware has also been identified at a site in the Tana River Valley along the eastern flank of Mt. Kenya (Merrick and Soper, n.d.). Thus, as is true of its temporal position, the geographic range of Kansyore ware remains poorly defined, but may be considerably more extensive than has previously been recognized (Collett and Robertshaw 1980; Robertshaw and Collett 1983a).

Narosura is basically a southern entity, occurring within and along the Gregory Rift from the Lake Naivasha basin in the north to the Lake Eyasi basin in the south, and extending up to about 80 km beyond the eastern and western margins of the Rift.

The southern limit of Maringishu ware more or less coincides with the northern boundary of Narosura ware. From there Maringishu ware extends northward in the Gregory Rift, occupying at least portions of the Lake Nakuru basin and the southern end of the Lake Bogoria (formerly Hannington) basin. The geographic relationship between Maringishu and Narosura wares leads me to suspect that the two entities occupied mutually exclusive areas at about the same time, but confirmation of this must await additional information on both the chronology and spatial distribution of Maringishu ware.

Elmenteitan ware is confined to the region between the Mau Escarpment, on the western margin of the Gregory Rift, and the eastern shore of Lake Nyanza. It thus lies basically west of the Maringishu area and northwest of the Narosura area.

Evolution of Stone Age Food-Producing Cultures in East Africa

As was mentioned earlier, Akira ware is very widely distributed in East Africa, within a spread that is, if anything, more extensive than that of Nderit ware. Sites with Akira occurrences are found in the Lake Turkana basin (Robbins 1980), on the Laikipia Highlands about 100 km northwest of Mt Kenya (Siiriainen 1977), at Lukenya Hill, (Bower *et al.* 1977) and in the Serengeti Plain (Bower 1973, Bower and Chadderdon 1986).

The Evolution of Pastoral Neolithic Cultures

In shifting from an essentially static to a dynamic perspective on PN cultures, one may begin by considering environmental factors that may have channeled their evolution. The most fundamental of these, and probably the most influential in the process at issue, is climate change, particularly increasing aridity during the early Holocene, which may have had the effect of forcing pastoralism out of the Sahara and attracting it into previously flooded portions of the Gregory Rift. Also relevant to this process was the drying out of swampy areas in intervening lands, such as the central Sudan (cf. Williams 1984:83). Thus the paleoenvironmental question that needs to be addressed here is, what is the chronology of Holocene moisture regimes in the regions of interest?

The Saharan record of Holocene climates is complicated by the fact that proxy data for wet/dry cycles sometimes exhibit poor chronological alignment in different parts of the region, perhaps because the cycles were time-transgressive (Williams, this volume) and also, no doubt, because of chronometric errors. Nevertheless, the general picture is reasonably clear. Moist conditions prevailed from about 11,000 to 4500 bp, except for perhaps three brief episodes of aridity lasting only a few hundred years each and centered respectively at about 8000 bp, 7500 bp and 6500 bp (McIntosh and McIntosh 1981; Wendorf *et al.* 1984:36-40; Williams 1984). While additional moist periods may have occurred after 4500 bp, the overall trend was apparently in the direction of modern hyperarid conditions. The general pattern for the Sahara was essentially paralleled in the climate history of Ethiopia and Somalia, where moist conditions during the early Holocene were interrupted by an episode (or episodes?) of aridity between about 8500 and 6500 bp and the trend toward a modern climatic regime was established some time between 5000 and 4000 bp (Brandt and Brook 1984; Brandt 1986; Gasse *et al.* 1980).

In East Africa the early Holocene was also a time of markedly increased moisture, such as to raise levels of lakes in the Gregory Rift, in some cases up to nearly 200 m above their present position (Butzer *et al.* 1972). Although land suitable for livestock herding may have been available in parts of the Rift that remained dry, as well as in areas outside the Rift, much of what is today prime grazing land in East Africa was inundated during large segments of Holocene time. Thus, a key paleonvironmental problem regarding the regional history of pastoralism is to determine the chronology of arid periods that resulted in major reduction of lake volume. Despite the availability of evidence for relatively dry episodes after about 7000 bp, the events are neither clearly correlated with arid periods to the north nor do they necessarily imply substantial exposure of flooded portions of the Rift. In fact, it is not until about 4000 bp that limnological evidence suggests a general retreat of East African lake levels to their present position (Richardson and Richardson 1972; Robertshaw, chapter 7). This is followed a few hundred years later by high elevation palynological evidence of a transition from a relatively moist climate to a regime resembling the modern one (Hamilton 1982).

Although dates for the earliest occurrence of domestic livestock in Africa are disputed (cf. Smith 1986), there is at least tenuous evidence that domestic cattle were present in the eastern Sahara by 8000 bp (Banks 1984; Wendorf et al. 1984). Since no wild progenitors for domestic cattle (or, for that matter, caprines) exist in contemporary or extinct East African faunas we may assume that the region's earliest domesticates were introduced from the north some time after 8000 bp. While Petit-Maire's evidence (this volume) from the Western Saharan suggests there may be a considerable time lag between the onset of climatic conditions favoring human occupation of marginally habitable regions and the appearance of substantial traces of a human presence, one would expect that major climatic deterioration in such areas would provoke rapid evacuation. Thus, on paleoenvironmental grounds the most likely dates for large-scale withdrawal of pastoral cultures from the Sahara were about 8000, 7500, 6500 and 4500 bp. (It is noteworthy that, of the three earlier arid episodes, the one at 7500 bp appears to have been the most intense). However, the earliest date for the establishment of livestock herding as a major subsistence regime in East Africa was probably some time after 4000 bp.

Various workers have called attention to pronounced variation in subsistence evidence among PN sites, whose faunal assemblages contain proportions of domestic elements that range from almost nil to predominant (Gifford et al. 1980; Gifford-Gonzalez 1984; Gifford-Gonzalez and Kimengich 1984; Nelson and Kimengich 1984). Equally heterogeneous are the data regarding the nature of PN settlements, with variables that relate to size, intensity and duration of occupation all ranging widely (Bower 1978; 1984a, b). Much of the diversity in PN subsistence and settlement evidence is probably geographically determined, that is, it may reflect functional differences among sites occupied by the same group or distinct subsistence-settlement systems practiced by diverse groups which, though belonging to the same general cultural entity, occupied widely separated and markedly different ecosystems. Both possibilities may be reflected in Table 6.2. Functional distinctions are implied by major differences in lithic and ceramic densities between two Nderit occurrences in the Serengeti National Park (HcJe1 and HbJd1) that, although separated by only 40 km, are geographically dissimilar. HcJe1 is situated in short-grass plains, along the route of the annual wildebeeste migration, but distant from a permanent source of water, while HbJd1 is located in woodlands near the Seronera River, a permanent stream. Differing from both HcJe1 and HbJd1 are the density data from GaJi4, an Nderit occurrence on the east side of Lake Turkana, more than 500 km north of the Serengeti sites in a generally more arid environment.

However, it is becoming increasingly evident that variation in PN subsistence and settlement practices may be a function of time, as well as geography. Thus, for example, Nderit occurrences at a given location tend to exhibit much lower artifact densities than later PN occupations at the same sites (compare, e.g., the Nederit and Akira occurrences at HcJe1 and HbJd1, Table 6.2). Moreover, sites representing the more recent entities generally contain more diverse inventories of material culture, including various relatively heavy items (e.g., stone bowls), than are found at Nderit sites. This is particularly true of Narosura and Elmenteitan occurrences, some of which also contain evidence of relatively persistent occupation, such as substantial midden accumulations and post molds (representing houses?). It is also noteworthy that the step-wise elaboration of material culture and intensification of settlement in later PN cultures seem to be paralleled by major social transformations as indicated by the appearance of evidence for complex mortuary ritual during the third millennium bp (M. D. Leakey 1945, 1966; Leakey and Leakey 1950; Siiriainen 1977).

These developments in cultural practices were apparently matched by changes in subsistence behavior, for the Nderit faunal assemblages are predominantly wild with a small fraction of domesticates (at HcJe1, exclusively *Bos taurus*: Gifford-Gonzalez personal communication; Bower and Smale n.d.), while the assemblages from sites representing more recent entities tend to be dominated by domesticates. This trend was by no means linear, since the fauna so far analyzed from Akira sites, representing the most recent PN entity, is predominantly wild (Gifford-Gonzalez and Kimengich 1984; Bower and Smale n.d.). Moreover, as I have indicated earlier, PN subsistence behavior (along with other aspects of culture) exhibits marked geographic, as well as temporal, variation. Nevertheless, present evidence suggests a major intensification of the role of domestic livestock in PN subsistence regimes during the third millennium bp, followed by a decline in consumption of the flesh of domestic animals during the second millennium bp (possibly involving a shift toward dairy pastoralism?).

Table 6.2: Variation in Lithic and Ceramic Densities Among PN Occurrences, Kenya and Tanzania

Sites	HcJe1[1]	HcJe1[1]	HbJd1[2]	HbJd1[2]	GaJi4[3]
Levels	(Nderit)	(Akira)	(Nderit)	(Akira)	(Nderit)
Lithic artifacts/m3	14,949	20,942	4,118	5623	223
Sherds/m3	39	151	200	860	44
Lithics:sherd	387:1	138:1	21:1	7:1	5:1

Note 1: Gol Kopjes site, Serengeti (Bower and Chadderdon 1986). Data from levels in which Nderit ware is abundant (20-40 cm b.s.) and levels dominated by Akira ware (10-15 cm b.s.).

Note 2: Seronera Game Lodge site (Bower 1973). Data from area SE-2, levels 3 (Nderit) and 2 (Akira).

Note 3: Data from Barthelme (1985). Values based on an average of maximum and minimum densities in a deposit whose thickness varies from 5 to 45 cm.

Conclusions

Despite the inadequacies of data regarding the Pastoral Neolithic, it seems clear that (a) a developmental stage consistent with its (revised) definition can be recognized in East Africa's archaeological record and (b) Wandibba's (1977) ceramic wares provide a reasonably secure framework for describing the culture history of the stage. It seems equally clear that, given the marked variability of PN subsistence-settlement practices and our limited knowledge of them, inferences concerning ecological adaptation and endogenous change in PN cultures can only be drawn in the most general terms. In closing this article, I would like to delineate broad patterns in the evolution of the PN and briefly compare them with those evident in certain other neolithic histories in sub-Saharan Africa.

The East African PN appears to pass through two sub-stages (see Table 6.3), the first of which was initiated some time earlier than 4000 bp, possibly as early as 7000 bp, if the Serengeti dates are correct. The earliest entity in this sub-stage seems to be Nderit ware, a ceramic tradition that is very widely distributed in East Africa but has not yet revealed any connection with ceramics from other regions. By about 4000 bp, several additional pottery wares are evident in sites scattered throughout the geographic range of the PN. Although little is known about the spatiotemporal distributions, faunal associations or antecedents of most of these entities, Kansyore ware is associated with domestic animals in at least one assemblage and is the only PN ware said to occur well to the north of Kenya and Uganda, in southern Sudan. In general, apart from the presence of domestic livestock, there is little to distinguish the first sub-stage of the PN from the LSA, in which it seems to be rooted.

The second sub-stage contains two phases, the first of which spans the third millennium bp, and is best represented by Narosura and Elmenteitan wares, though other entities are likely to be recognized in future (cf. Hivernel 1983). There is no evidence of typological continuity between wares of this phase and any of the first sub-stage wares. Although there are no published claims of northerly connections for wares representing the early phase of the second sub-stage, I have seen sherds from Quiha rockshelter (Clark 1954) near Lake Tana (Ethiopia), that closely resemble Narosura ware. However, additional evidence is required to confirm this link. The early phase of the second sub-stage is clearly an evolved Neolithic, whose cultures relied heavily upon domesticates and exhibited a high degree of technical and social elaboration.

The late phase of the second sub-stage was relatively brief, lasting only approximately 600 years, from about 1900 to 1300 bp. It is represented by Akira ware—possibly also Maringishu ware, though the temporal position of the latter entity is unclear. While the first sub-stage and the early phase of the second are sharply distinguished from one another by formal and chronological boundaries, the two phases of the second sub-stage tend to merge typologically and overlap temporally. Thus, as various observers have noted, there are close resemblances in decorative motif and/or technique between Akira and Narosura wares, as well as between the latter and Maringishu ware (Bower *et al.* 1977; Wandibba 1977; Collett and Robertshaw 1983). However, the subsistence regimes of the two phases seem to differ substantially (Table 6.3).

Three aspects of the sequence outlined above are particularly noteworthy: the apparent inertia of the first sub-stage, the considerable evidence of discontinuity between the first and second substages and nonlinear trends in subsistence regime. Let me comment briefly on each of these.

The static character of the first PN sub-stage depends on two considerations: the sub-stage's duration and the degree of culture change it exhibits. As to the latter, the only new trait of consequence in early PN cultures is domestic animals, since pottery may have been made in East Africa before livestock were introduced (e.g., Phillipson

1985:141; Robbins 1972). It is, of course, possible that major cultural transformations occurred without leaving a recognizable archaeological expression. However, the inference of stasis is strengthened by evidence of continuity in lithic traditions with the preceding LSA, as noted by Ambrose (1984) and also apparent in the Serengeti sequence (personal observation).

Table 6.3: Major Features of PN Evolution

Sub Stages	Dates (bp)	Wares	Meat Consumption
Second Sub-Stage, late phase	1300	(others?) Maringishu (?)	Wild fauna, some domestic animals
	1900	Akira	
Second Sub-Stage, early phase		(others?) Elmenteitan	Predominantly domestic animals
	3000	Narosura	
First Sub-Stage		(others?) Ileret Salasun Kansyore	Wild fauna, some domestic animals
	7000 (?)	Nderit	

The question as to how long the first sub-stage lasted is vexed by differing views as to the earliest dates for Nderit ware, which range from somewhat earlier than 7000 bp (as argued here) to approximately 4000 bp (cf. Phillipson 1977, 1982; Barthelme 1985). But, even if the conventional (i.e., short) chronology for Nderit ware should prove correct, we are still left with a situation in which, for upwards of a millennium, the presence of domestic animals in East Africa had little appreciable effect on cultures of the region. Such a situation cries out for comparative study.

The discontinuity in ceramic typology between the first and second sub-stages of the PN suggests cultural replacement, perhaps as a result of immigration from the north, while the marked shift in broad patterns of subsistence behavior between the sub-stages implies a basic change in adaptive strategy. In fact, the interpretation of the second sub-stage as a new adaptation to regional geography may be valid regardless of whether the adaptation involved autochthonous development or cultural replacement. However, one important difficulty with this interpretation is that the cultural changes implied by the onset of the second sub-stage do not coincide with any major environmental transformation so far recognized in the late Holocene of East Africa. Indeed, the second sub-stage seems to have begun more or less 1000 years after the lakes of the Gregory Rift had shrunk to about their modern levels. If this paleoenvironmental development was, nevertheless, one of the causes of the change in adaptive strategy that characterizes the second sub-stage, the apparent lag between environmental shift and cultural response may parallel the one observed by Petit-Maire (this volume) in the western Sahara.

Although the difference in subsistence regime between the early and late phases of the PN's second sub-stage evident in Table 6.3 are based on faunal data from a very limited number of sites, it seems likely that some kind of change in adaptive strategy separates the two phases. This is implied by differences in equipment and settlement; for example, Akira pottery seems more specialized than earlier wares, is associated with fewer heavy ground stone objects, such as stone bowls and platters, and occurs at sites that lack evidence of house construction, such as postmolds. But the changes in adaptive strategy in this case seem to indicate an autochthonous development, since the ceramic wares of the two phases are linked by similarities in decorative technique and motif.

Setting aside the question as to the mechanism for PN cultural transformations (whether by autochthonous development, stimulus diffusion or migration), it is clear that, from an economic point of view, the evolution of the PN is distinctly nonlinear. That is, the history of subsistence regimes does not reflect a progressive "neolithization," involving a steadily increasing reliance upon livestock. Instead, there is a step-wise shift toward the consumption of domestic fauna in the early phase of the second substage, followed by a return to a greater emphasis on wild meat in the late phase. These are, of course, very broad and rather tenuous generalizations that ignore evidence of substantial variation in PN subsistence patterns at any given time. Taking the generalizations at face value, however, one may argue that either or both subsistence patterns of the second substage constitute ecological adaptations. But it is difficult to see any adaptive significance in the meager presence of domestic livestock during the first sub-stage of the PN.

Assuming that the broad characteristics of the PN discussed in the last few paragraphs are substantiated by future research, the "bow wave" image of the spread of food production into East Africa will need to be replaced by a more complex model. In this connection, it is useful to look at neolithic histories in other regions, particularly the Nile Valley and West Africa.

The Khartoum Neolithic, as represented at sites such as Kadero (Krzyzaniak 1978), seems to have emerged within at most a few centuries as a highly evolved, early food-producing cultural stage, with an abundant material inventory (e.g., vast quantities of pottery and ground stone implements), elaborate mortuary ritual, large quantities of domestic animals, intensively occupied sites, etc. Such explosive development was evidently characteristic of the neolithic throughout the Nile Valley, from Khartoum to the delta between about 6000 and 5500 bp. (Hassan 1986). This situation, which stands in marked contrast with the very slow development of the East African PN, may reflect a direct response to ecological pressures in nearby regions.

Since livestock, which are the *sine qua non* of the PN, were domesticated in areas remote from East Africa, the origins of the stage must in some measure, be connected with processes that unfolded well over 1000 kilometers to the north. The chain of events that led to the introduction of livestock in East Africa was attenuated not only by distance but also by having passed through such ecologically stressful regions as southern Sudan and Ethiopia, which may have been tsetse infested, (cf. Smith 1984) and the filtering effect of zones of contact between food-producers and hunter-gatherers (cf. Bogucki 1987), perhaps not altogether different from the effects of national boundaries upon drought refugees in contemporary Africa.

Although the data needed for tracing this long and complicated chain of events are presently unavailable, it is difficult to imagine that the process will turn out to have consisted merely of a southward migration of pastoralists responding to the push-pull of environmental deterioration in the north and the expansion of grasslands in East Africa. To return to a physical metaphor, it seems much more likely that the process can best be described in terms of "trickle and splash" rather than a "bow wave" of advancing pastoralism. Such a "trickle and splash" model seems to be particularly well suited to the first stage of the East African PN, which is almost certainly ultimately connected with an early-mid Holocene episode of aridity in the Sahara. Present evidence from the Serengeti suggests that the arid period involved may have been the one centering at about 7500 bp. It also suggests that the earliest PN cultural entities catalyzed by the "trickle and splash" of livestock into East Africa may have spread from south to north, as apparently was the case with the early food-producing Kimtampo culture of West Africa (Stahl 1985).

As regards the evolution of the PN, while it is likely that the decline of the Gregory Rift lakes at about 4000 bp provided opportunities for the expansion of herding cultures, no direct causal link can be established between the drying-up of the lakes and the onset of the second PN substage, since the two events were temporally separated by about 1000 years. In the present state of knowledge, it would be premature to rule out an ecological explanation for the emergence of the mature PN. However, as Robertshaw has suggested (this volume), it is also necessary to consider the possibility that social processes may have played an important role in the development of a pastoral life-way in East Africa. At a more general level, it seems likely to me that the entire history of the relationship between human cultures and domestic livestock in the region has been shaped largely by social factors upon which environmental opportunities and constraints have had complex, subtle and sometimes negligible effects.

Acknowledgements

Research results reported in this chapter were obtained with the support of grants awarded to the author by the Iowa State University Research Foundation, the L. S. B. Leakey Foundation, and the National Science Foundation (grant #BNS 82-19416). All research accomplished in Tanzania was done in collaboration with the Tanzania Antiquities Division, whose contributions are gratefully acknowledged.

References

Ambrose, S. H. 1984a. The introduction of pastoral adaptations to the highlands of East Africa. In *From Hunters to Farmers* (eds J. D. Clark and S. A. Brandt): pp. 212–239. Berkeley: University of California Press.

Ambrose, S. H. 1984b. *Holocene environments and human adaptations in the central Rift Valley, Kenya.* (Ph.D. thesis, University of California, Berkeley).

Banks, K. M. 1984. *Climates, Cultures and Cattle: The Holocene Archaeology of the Eastern Sahara.* Dallas: Department of Anthropology, Institute for the Study of Earth and Man, Southern Methodist University.

_____. 1984. Early evidence for animal domestication in eastern Africa. In *From Hunters to Farmers* (eds J. D. Clark and S. A. Brandt): pp. 200–5. Berkeley: University of California Press.

Barthelme, J. W. 1985. *Fisher-hunters and Neolithic Pastoralists in East Turkana, Kenya*. Oxford: B.A.R. International Series 254 (Cambridge Monographs in African Archaeology 13).

Bishop, W. W. and Clark, J. D. (eds) 1967. *Background to Evolution in Africa*. Chicago: The University of Chicago Press.

Bogucki, P. 1987. The establishment of agrarian communities on the North European Plain. *Current Anthropology* 28(1):1–24.

Bower, J. 1973. Seronera: excavations at a Stone Bowl site in the Serengeti National Park, Tanzania. *Azania* 8:71–104.

_____. 1978. Culture, environment and technology: preliminary results of an archaeological study in Kenya. *Proc. Ia. Acad. Sci.* 85(2):41–44.

_____. 1984a. Settlement behavior of pastoral cultures in East Africa. In *From Hunters to Farmers* (eds J. D. Clark and S. A. Brandt): pp. 252–260. Berkeley: University of California Press.

_____. 1984b. Subsistence-settlement systems of the Pastoral Neolithic in East Africa. In *Origin and Early Development of Food-Producing Cultures in North-Eastern Africa* (eds L. Krzyzaniak and M. Kobusiewicz): pp. 473–480. Poznan: Polish Academy of Sciences and Poznan Archaeological Museum.

_____., and Chadderdon, T. J. 1986. Excavation of Pastoral Neolithic sites in the Serengeti National Park, Tanzania. *Azania* (in press).

_____., Nelson, C. M., Waibel, A. F. and Wandibba, S. 1977. The University of Massachusetts' Later Stone Age/Pastoral Neolithic comparative study in central Kenya: an overview. *Azania* 12:119–46.

_____. and Nelson, C. M. 1978. Early pottery and pastoral cultures of the central Rift Valley, Kenya. *Man* 13 (n.s.): 554–66.

_____. and Smale, W. In preparation. Emergence of the Pastoral Neolithic in the Serengeti Plains, Tanzania.

Brandt, S. A. 1986. The Upper Pleistocene and Holocene prehistory of the Horn of Africa. *African Archaeological Review* 4:41–82.

_____. and Brook, G. A. 1984. Archaeological and paleoenvironmental research in northern Somalia. *Current Anthropology* 25:119–21.

Butzer, K. W., Isaac, G. Ll., Richardson, J. L. and Washbourn-Kamau, C. 1972. Radiocarbon dating of East African lake levels. *Science* 175:1069–76.

Clark, J. D. 1954. *The Prehistoric Cultures of the Horn of Africa.* Cambridge: Cambridge University Press.

_____., Cole, G. H., Isaac, G. L. and Kleindienst, M. R. 1966. Precision and definition in African archaeology. *S. Afr. Archaeol. Bull.* 21:114–121.

Cohen, M. 1970. A re-assessment of the Stone Bowl Cultures of the Rift Valley, Kenya. *Azania* 5:27–38.

Cole, S. 1963. *The Prehistory of East Africa* New York: Mentor Books.

Collett, D. P. and Robertshaw, P. T. 1980. Early Iron Age and Kansyore Pottery: finds from Gogo Falls, South Nyanza. *Azania* 15:133–145.

_____. 1983. Pottery traditions of early pastoral communities in Kenya. *Azania* 18:107–126.

Gasse, F., Rognon, R. and Street, F. A. 1980. Quaternary history of the Afar and Ethiopian Rift lakes. In *The Sahara and the Nile* (eds M. A. J. Williams and H. Faure): pp. 361–400. Rotterdam: A. A. Balkema.

Gifford, D. P., Isaac, G. L. and Nelson, C. M. 1980. Evidence for predation and pastoralism at Prolonged Drift: a Pastoral Neolithic site in Kenya. *Azania* 15: 57–108.

Gifford-Gonzalez, D. P. 1984. Implications of a faunal assemblage from a Pastoral Neolithic site in Kenya: findings and a perspective on research. In *From Hunters to Farmers* (eds J. D. Clark and S. A. Brandt): pp. 240–251. Berkeley: University of California Press.

Gifford-Gonzalez, D. P. and Kimengich, J. 1984. Faunal evidence for early stock-keeping in the Central Rift of Kenya: preliminary findings. In *Origin and Early Development of Food-Producing Cultures in North-Eastern Africa* (eds L. Krzyzaniak and M. Kobusiewicz): pp. 457–471. Poznan: Polish Academy of Sciences and Poznan Archaeological Museum.

Gramly, R. M. 1975. *Pastoralists and hunters: recent prehistory in Southern Kenya and Northern Tanzania.* (Ph.D. thesis, Harvard University).

Hamilton, A. C. 1982. *Environmental History of East Africa.* London: Academic Press.

Hassan, F. A. 1986. Chronology of the Khartoum "Mesolithic" and "Neolithic" and related sites in the Sudan: statistical analysis and comparisons with Egypt. *African Archaeological Review* 4:83–102.

Hivernel, F. 1983. Excavations at Ngenyn (Baringo District, Kenya). *Azania* 18:45–79.

Krzyzaniak, L. 1978. New light on early food-production in the central Sudan. *Journal of African History* 19:159–72.

Leakey, L. S. B. 1931. *The Stone-Age Cultures of Kenya Colony*. Cambridge: Cambridge University Press.

Leakey, M. D. 1945. Report on the excavations of Hyrax Hill, Nakuru, Kenya Colony, 1937–38. *Trans. Roy. Soc. S. Afr.* 30(4):271–409.

———. 1966. Excavation of burial mounds in Ngorongoro Crater. *Tanzania Notes and Records* 66:1–13.

Leakey, M. D. and Leakey, L. S. B. 1950. *Excavations at Njoro River Cave*. Oxford: Oxford University Press.

McIntosh, S. K. and McIntosh, R. J. 1981. West African prehistory. *American Scientist* 69:602–13.

Mehlman, M. J. 1977. Excavations at Nasera Rock, Tanzania. *Azania* 12:111–18.

Melman, M. J. 1979. Mumba-Hohle revisited: the relevance of a forgotten excavation to some current issues in East African prehistory. *World Archaeology* 11:80–94.

Merrick, H. V., and Soper, R. n.d. Kiambere Hydro-Electric Power Project Pre-Construction Environmental Impact Study (unpublished report).

Nelson, C. M., and Kimengich, J. 1984. Early phases of pastoral adaptation in the central highlands of Kenya. In *Origin and Early Development of Food-Producing Cultures in North-Eastern Africa* (eds L. Krzyzaniak and M. Kobusiewicz): pp. 481–487. Poznan: Polish Academy of Sciences and Poznan Archaeological Museum.

Odner, K. 1972. Excavations at Narosura, a stone bowl site in the southern Kenya highlands. *Azania* 7:25–92.

Onyango-Abuje, J. C. 1977. *A contribution to the study of the Neolithic in East Africa with particular reference to the Naivasha-Nakuru Basin*. (Ph.D. thesis, University of California, Berkeley).

Phillipson, D. W. 1977. *The Later Prehistory of Eastern and Southern Africa*. London: Heinemann.

Phillipson, D. W. 1982. Early food production in sub-Saharan Africa. In *The Cambridge History of Africa*, Volume 1 (ed J. D. Clark): pp. 770–829. Cambridge: Cambridge University Press.

Phillipson, D. W. 1985. *African Archaeology*. Cambridge: Cambridge University Press.

Richardson, J. L. and Richardson, A. E. 1972. History of an African rift lake and its climatic implications. *Ecological Monographs* 42:449–534.

Robbins, L. H. 1972. Archaeology in Turkana District, Kenya. *Science* 176:359–366.

_____. 1980. *Lopoy: a Late Stone-Age Fishing and Pastoralist Settlement in the Lake Turkana Basin, Kenya*. Michigan State University, Anthropological Series, Volume 3, No. 1.

Robbins, L. H., McFarlin, S. A., Brower, J. L. and Hoffman, A. E. 1977. Rangi: A Late Stone Age site in Karamoja District, Uganda. *Azania* 12:209–233.

Robertshaw, P. and Collett, D. 1983a. A new framework for the study of early pastoral communities in East Africa. *Journal of African History* 24:289–301.

Robertshaw, P. and Collett, D. 1983b. The identification of pastoral peoples in the archaeological record: an example from East Africa. *World Archaeology* 15:67–78.

Robertshaw, P., Collett, D., Gifford, D. and Mbae, N. B. 1983. Shell middens on the shores of Lake Victoria. *Azania* 18:1–43.

Sackett, J. R. 1982. Approaches to style in lithic archaeology. *Journal of Anthropological Archaeology* 1:59–112.

Siiriainen, A. 1977. Later Stone Age in the Laikipia Highlands. *Azania* 12:161–86.

Smith, A. B. 1984. Environmental limitations on prehistoric pastoralism in Africa. *African Archaeological Review* 2:99–111.

Smith, A. B. 1986. Cattle domestication in North Africa. *African Archaeological Review* 4:197–203.

Soper, R. C. and Golden, B. 1969. An archaeological survey of Mwanza region, Tanzania. *Azania* 4:15–79.

Tolstoy, P. 1987. Transpacific echoes and resonances. *Quarterly Review of Archaeology* 7(3–4).

Stahl, A. B. 1985. Reinvestigation of Kintampo 6 rock shelter, Ghana: implications for the nature of culture change. *African Archaeological Review* 3:117–50.

Wandibba, S. 1977. *An attribute analysis of the ceramics of the Early Pastoralist Period from the Southern Rift Valley, Kenya.* (M. A. thesis, University of Nairobi).

Wendorf, F. Schild, R. and Close, A. E. 1984. *Cattle-Keepers of the Eastern Sahara: The Neolithic of Bir Kiseiba.* Dallas: Department of Anthropology, Institute for the Study of Earth and Man, Southern Methodist University.

Willey, G. R. and Phillips, P. 1958. *Method and Theory in American Archaeology.* Chicago: University of Chicago Press.

Williams, M. A. J. 1984. Late Quaternary prehistoric environments in the Sahara. In *From Hunters to Farmers* (eds J. D. Clark and S. A. Brandt): pp. 74–83. Berkeley: University of California Press.

Evolution of Stone Age Food-Producing Cultures in East Africa

Environment and Culture in the Late Quaternary of Eastern Africa: A Critique of Some Correlations

By Peter Robertshaw, British Institute in Eastern Africa, Nairobi.

The last two decades have seen a considerable amount of research into the environmental history of the late Quaternary of East Africa. Studies of lake levels (e.g., Butzer *et al.* 1972; Gasse 1980), pollen (e.g., Coetzee 1967; Hamilton 1982), and the former extents of glaciers (e.g., Osmaston 1975) have provided archaeologists with a rich, if complex and not always entirely consistent, body of paleoenvironmental information against which to consider the results of excavations. Indeed there has been a close link between reconstructions of prehistory and past environments since archaeology began in East Africa; for example, Leakey (1931) linked the prehistoric cultures that he identified to a sequence of climatic episodes, and argued that hunter-gatherer settlement patterns would have varied between pluvial and interpluvial periods. However, it should be noted that much of Leakey's concern with prehistoric climates in East Africa was aimed at the establishment of correlations with the sequence of European ice ages "so to learn much concerning the comparative dating of the various Stone Ages in these two continents" (Leakey 1931:8; see also Wayland 1924).

In recent years, archaeologists working at the *synchronic* level have tended to *correlate* their archaeological findings with paleoenvironmental data; for example, evidence for fishing at 9000 bp has been correlated with high lake levels. On the *diachronic* level changes observed in the archaeological record have been *"explained"* by reference to environmental changes. The synchronic correlation between environment and culture is generally reached through the mediation of the concept of *adaptation*. For example, we might read in almost any work of prehistory that "during the X cold period the A people were adapted to the hunting of the large game animals which roamed the vast grass plains." Thus "adaptation" implies or involves subsistence which in turn may be linked to technology, settlement patterns, and a whole host of other traits, such as high lake levels, fishing, harpoons, semi-permanent settlements, egalitarianism, monogamy, etc. However, correlating environment and culture by invoking adaptation does not tell us anything about how or why any particular cultural manifestation came into being. As has been said before, "the concept of adaptation, unless combined with a demonstrable principle of selection, loses all significance: without it, anything that exists and therefore functions is, *ipso facto*, adapted" (Godelier 1972:xxxiv, cited by Ingold 1983:14). Thus, to move from the domain of description to the domain of explanation we must switch our focus from the synchronic to the diachronic.

Environment and Culture in the Late Quaternary of Eastern Africa

Cultural change in African prehistory has both often, and indeed generally, been considered to be caused ultimately by environmental change. Why this should be so lies outside the scope of the present chapter and requires detailed historiographical treatment. The argument for the environmental causation of cultural change runs roughly as follows: vegetation changes resulting from a climatic shift lead to a reduction in the abundance and availability of the resources exploited by a prehistoric group with the result that to avoid starvation and population decline the group must shift its attention to other resources or move elsewhere. If the first option is chosen, then adaptation to the new configuration of resources is likely to involve changes in technology, subsistence strategies, settlement patterns, and probably other variables, which in sum comprise cultural change as understood by most archaeologists. If the second option, that of movement elsewhere, is chosen, cultural change is also likely to result from adaptation to slightly different surroundings including the preexisting human population of the newly settled region. In both cases then the ultimate cause of cultural change is seen as environmental change. If we reverse the sequence of the argument and perhaps state it the way most archaeologists might implicitly conceive it, then any cultural change noted in the archaeological record is assumed to be the result of a changing adaptation, and for any adaptive shift there must be an environmental change to precipitate it. Stated this latter way we have arrived at an explanation of culture change which verges on environmental determinism.

This deterministic approach to culture change fails to consider other nonenvironmental variables which might equally account for changes observed in the archaeological record. Indeed one can perhaps discern an attitude among some archaeologists that explanations of culture change which do not invoke environmental change are in some manner nonfalsifiable or even nonscientific. The remainder of this chapter examines two case studies from the later prehistory of Eastern Africa in which environmental-cultural correlations and explanations have been attempted. The details of the environmental-cultural correlations are shown to be problematic and possibly misconstrued in each case. The reasons for this include failure to consider sampling biases in the archaeological record, misreadings, or misunderstandings of the environmental evidence, and more generally the paucity of the available archaeological information. Alternative explanations involving both environmental and other variables are tentatively outlined. The exercise has two purposes: first, to show that archaeologists need to be both rigorous and cautious in their application of paleoenvironmental data to archaeological problems, while remembering the limitations of the data from their own discipline; and, second, to draw attention to the

P. Robertshaw

explanatory potential offered by careful consideration of factors which may bear no direct relation to environmental variables.

Case 1: The Development and Spread of Microlithic Technology

It has been argued that "the adoption of backed microlithic technology in many areas (not only in sub-Saharan Africa but also in much of Eurasia) was initially linked to the development of new hunting techniques to permit the exploitation of more densely vegetated environments" (Phillipson 1980a:15). Thus it is suggested that adaptation to an environmental shift towards an increased density of vegetational cover involved an increased reliance on the bow and microlith-tipped arrow, which were particularly effective for hunting the small solitary game of the expanding woodlands (Phillipson 1977:31; 1976:201). However, since microlithic industries are not found exclusively in wooded environments, the use of the bow and arrow is thought to have spread later into different environments from that in which it developed (Phillipson 1980b:230). Phillipson's hypothesis is comprised of two arguments: 1) that backed microliths, but not necessarily all of them, were hafted to form the barbs of points of arrows; 2) that the appearance of microlithic technology can be correlated with an environmental change to more wooded vegetation, due presumably to a warmer and wetter climate than existed previously.

The suggestion that some microliths were hafted in arrows receives somewhat tentative support from the study of traces of mastic found on specimens in eastern Zambia (Phillipson 1976). Historic sources show that microliths were certainly used in arrows in the period of European contact in Southern Africa (Clark 1977; Rudner 1979). However, the microliths=hunting stereotype has been criticised by David Clarke (1976), who has shown that a wide variety of composite tools with large numbers of microliths are documented ethnographically as being used in plant-gathering, harvesting, and a great number of processing activities. Furthermore, microlithic industries are found in Australia, where there is no evidence from ethnography of Aboriginal use of bows and arrows. Studies of use-wear combined with experimental work have indicated that Later Stone Age microliths excavated in Ethiopia may have been used in processing plant materials, including grass-cutting (Clark and Prince 1978). Therefore, one could argue, contrary to Phillipson, that the adoption of microlithic technology was linked to the increased importance of the exploitation of plant materials for both food and artefacts. The use of microliths in arrows may have been of subsidiary importance and may have occurred much later. The microliths=plant processing correlation certainly accounts better for the presence of nonbacked microliths such as scrapers and adzes, in addition to backed microliths, in the earliest microlithic industries;

these tool types could not possibly have formed the barbs or points of arrows. However, as Parkington has pointed out, there is a basic fallacy in both the microliths=hunting and microliths=gathering correlations when they are extended to become environmental correlates. For it is not microlithic technology which has environmental correlates, but interassemblage differences within the technology. The technology itself is simply "a pool of technological and typological resources into which prehistoric people...dipped in order to carry out tasks of various sorts" (Parkington 1979:74).

Early microlithic industries in Eastern Africa have been found at Matupi in northeastern Zaire *ca.* 40,000 bp (Van Noten 1982), at Lake Besaka in Ethiopia (Brandt 1982) *ca.* 22,000 bp, at Lukenya Hill in Kenya *ca.* 20,000 bp (Merrick 1975; Gramly 1976; Miller 1979), at Nasera in Tanzania *ca.* 21,500 bp (Mehlman personal communication), at Kisese in Tanzania *ca.* 19,000 bp (Inskeep 1962), at Olduvai *ca.* 17,000 bp (Leakey *et al.* 1972), at Munyama on Lake Victoria *ca.* 16,000 bp (Van Noten 1971), and in various "Nachikufan I" sites in Zambia *ca.* 17,000 bp (Phillipson 1976). Unfortunately, faunal remains are preserved in few sites, and rarely reported in the literature. The argument for a shift to the hunting of animals from a more wooded environment is barely supported by the tiny sample of faunal remains from the Nachikufan I levels at Kalemba in which zebra and wildebeest are still represented (Phillipson 1976:Table 51). Similarly, the presence of hartebeest, gazelle, and zebra in the Lukenya Hill fauna indicates open savanna (Miller 1979:31). Only at Lake Besaka and Matupi does there appear to be good evidence for relatively closed vegetation, but even the latter site was probably situated in savanna with closed forest nearby (Van Noten 1982:32). In addition, much of the evidence from lake levels and pollen cores indicates cold, dry, open country conditions from *ca.* 20,000 to *ca.* 12,000 bp, though Lake Mobutu may have had high levels between 18,000 and 14,000 bp and the level of Lake Manyara was somewhat higher between 19,500 and 16,500 bp (Gasse 1980).

Thus the archaeological and paleoenvironmental evidence from Eastern Africa in support of Phillipson's hypothesis is equivocal. However, in the Southern Cape the first microlithic industry, the Robberg, is associated with a fauna indicative of open grassland (Deacon 1980). Given these data and the critique of the microliths=hunting correlation, it seems clear that the hypothesis linking the appearance of backed microliths with the exploitation of densely vegetated environments is likely to be rejected, at least in its application to much of sub-Saharan Africa. Two alternative hypotheses, which may be interdependent, can be suggested and there are no doubt many other hypotheses that could be formulated.

The first alternative has been mooted by Parkington (1979:74): "the near universality of...[microlithic industries] in the terminal Pleistocene and Holocene...suggests that the dispersion has more to do with an increased awareness of technological possibilities than with similar adaptations to environmental change." Thus Parkington's hypothesis rests on the idea of a mental or conceptual breakthrough by prehistoric peoples in the late Pleistocene.

The second alternative hypothesis, which is not incompatible with the first, is that the development of microlithic technology occurred in conjunction with an expansion of the food base. In particular, there may have been increasing emphasis on plant foods, the gathering and processing of which microlithic tools and technology may have facilitated. Widening of the range of exploited resources in Eastern and Southern Africa may be part of or analogous to the "broad spectrum revolution" originally proposed by Flannery (1969) for the Near East. In the Near East and Europe, this "revolution" resulted in increased emphasis on subsistence from aquatic resources, small game, and plant foods. The same shift in subsistence may also have occurred in sub-Saharan Africa; however, food remains other than mammalian bones rarely survive from this period. Nevertheless, one may note the discovery of shell middens at Mumba-Höhle associated with a stone artefact assemblage with characteristics transitional from the MSA to the LSA (Mehlman 1979). Broadening of the subsistence base may have been a response to increased human population density causing a "squeeze" in some regions on the availability of larger game (Binford 1983:212-3). Therefore, microlithic technology may have been developed in Eastern Africa to provide the variety of tools necessary for the exploitation of a much greater range of resources than was previously attempted. This dietary change may have been an adjustment to human population densities in some regions reaching the carrying capacity permitted by the hunting of larger game. No environmental change needs to be posited.

Sceptics might well make the accusation that this second alternative hypothesis simply represents a substitution of a form of demographic determinism in place of the environmental causation that has been criticised above. While this charge may contain an element of truth, it does not reduce the potential explanatory value of the hypothesis nor does it detract from the general point that not all cultural change need be linked to environmental change. All the models discussed here are in fact probably simplistic; if the data indicate anything it is that the appearance of microlithic technology is a complex phenomenon for which any single factor explanation is likely to prove inadequate. It is, for example, relatively clear that the manufacture of the first backed microliths in Eastern Africa is preceded by a period, probably of several millennia or more, in which microlithic technology was used for the making of *outils écaillés*, small scrapers, burins, and other tool forms. This is attested not only by the tools themselves but also by the presence of bipolar cores and flaking debris. Furthermore, the origins of microlithic technology in sub-Saharan Africa cannot be ascribed to a movement of people or diffusion of ideas from north of the Sahara. Rather it appears that its roots lie in the local Middle Stone Age and that for most of the region there is no intervening period corresponding with the blade industries of the European Upper Paleolithic. This is, however, not to deny that some diffusion of the knowledge of microlithic technology and backed microlith manufacture may have occurred within sub-Saharan Africa; indeed the dating of the earliest backed microliths at many sites in Eastern Africa to around 20,000 bp is suggestive of this process.

Case 2: The spread of pastoralism to the East African Highlands

Various hypotheses ranging from independent domestication in East Africa, through diffusion, to small-scale or large-scale population movements have been suggested to explain how pastoralism spread to the highland areas of East Africa, i.e., the Central Rift Valley and nearby savanna grasslands. The most recent hypothesis in which climatic and environmental changes play important roles has been put forward by Ambrose (1984a; 1982). It is suggested that population pressure among mixed farmers in Southern Ethiopia caused various groups to bud off between 5000 and 4500 bp into the northern Kenya lowlands where the arid conditions favoured pastoral subsistence. However, pastoralists were unable to settle in central Kenya until around 3000 bp. The reason for this is an arid climate phase in the East African highlands from 5600 to 3000 bp in which the vegetation shifted towards arid bush, possibly harbouring tsetse fly. Thus only with climatic amelioration towards modern conditions around 3000 bp were pastoralists able to move south (Ambrose 1982:138; 1984a:228, 235–6). Before discussing this hypothesis it should immediately be noted that relevant archaeological observations are generally few and far between. Thus, almost nothing is known of the prehistory of Southern Ethiopia or of the region between Lake Turkana and the Central Rift. Even within the highlands excavations have tended to be small scale and publication beyond the level of preliminary reports a rare occurrence. Therefore, every hypothesis concerning the spread of pastoralism is likely to be speculative to a considerable degree; adequate testing of competing hypotheses will require a great deal more field work and analysis. Thus, no claims of veracity are made for the preliminary and tentative scenario of the spread of pastoralism outlined below. The purpose of the exercise is more to draw attention to possible sampling biases in the archaeological and paleoenvironmental record and to indicate where paleoenvironmental evidence may have been misread than to imply that the alternative model proposed is correct.

The suggestion of an arid climatic phase in the Central Rift between 5,600 and 3,000 bp is based on the evidence obtained from cores drilled in Lake Naivasha (Richardson and Richardson 1972). However, the interpretation of this evidence which would argue that the climate "became more arid than at present" turning "the open savanna into tsetse-infested bush" (Ambrose 1982:138) may be an exaggeration.[1] Richardson and Richardson (1972:529–530) wrote that during the period 5650-4000 bp the lake "fell to levels similar to today's" and during this decline "the climate of the Naivasha basin *may* have been somewhat drier than at present; however, *there is no way to be sure of this*" (emphasis added). The climate between 4000 and 3000 bp was "rather stable," though the part of the lake that was cored dried out at about 3000 bp for less than a hundred years. However, "we doubt if yearly rainfall at that time [3000 bp] was markedly different from today's" (Richardson and Richardson 1972:530). Temperatures may have been slightly higher than those of today and a more seasonal pattern of precipitation may have persisted until about 2000 bp. Thus it is far from clear whether these slight climatic changes would have any pronounced effect on the vegetation of the region. Certainly pastoralists live today and have done so in the past in regions far more arid than the Central Rift; thus they can only have been excluded from the latter region before 3000 bp by a very extensive tsetse fly belt.

Most of the Central Rift is tsetse free today and seems to have been so throughout the historic period (*Survey of Kenya* 1970). Flies of the *morsitans*-group are generally found in areas with temperatures between 20° and 28° C, relative humidity between 50 and 80 percent, and rainfall between 25 and 60 inches (Lambrecht 1964:10). Naivasha today is close to the lower rainfall limit; thus any tendency towards less rainfall and higher temperatures reducing humidity would make the region even more marginal for tsetse. Therefore, though flies may have been able to survive and spread in the rainy season in the period between 4000 and 3,000 bp, the long dry season would probably have restricted their range to heavily vegetated water courses. Such a pattern is found in parts of southwestern Kenya today. Thus, it is difficult to believe that

---

[1] During the discussion at the symposium which followed the presentation of this paper Stanley Ambrose kindly presented aspects of his reconstruction of the Holocene environments of the Central Rift which are more complex and supersede his earlier published statements (see also Ambrose 1984b:33–4). It has not been possible to take account of this new information here.

trypanosomiasis would have been a barrier to the spread of pastoralists into the Central Rift.

While this critique of the climatic and paleoenvironmental scenario that has been suggested for the Central Rift is not necessarily more accurate than the original version, it does show that paleoclimatic evidence is as open to reconsideration as archaeological data. However, rather more disturbing are the results from a core drilled in Lake Naivasha more recently. These indicate that the major wet phase continued until 4145 bp when conditions like those of the present seem to have been rapidly established (Richardson cited personal communication in Kamau 1973). The evidence from this new core would seem to match better than the earlier cores the geomorphological results from Lake Nakuru (Butzer et al. 1972). Pollen spectra collected in various mountains also document a shift from wetter conditions to dry conditions similar to those of the present around 3700 bp (Hamilton 1982). Therefore, the hypothesis of a climatic episode more arid than that of the present between 5600 and 3000 bp in the Central Rift region is open to reevaluation,

A careful review of the dating evidence from the Turkana Basin indicates that the presence of pastoralists is not certainly established until about 4,000 bp. (Collett and Robertshaw 1983a). Given the close similarity between the pottery from the sites dated to 4000 bp in the Turkana region and pottery from undated sites like Stable's Drift in the Central Rift (Collett and Robertshaw 1983b), it seems reasonable to suggest that evidence for pastoralists in the latter region soon after 4000 bp will eventually be forthcoming. Thus, there may have been no long hiatus between pastoralist occupation of the Turkana basin and their expansion into the Central Rift region. Whether this expansion may be linked to the evidence for climatic change suggested around 3700 bp is a matter for further research.

Examination of the pottery assemblages of the early pastoralist communities of Kenya (Collett and Robertshaw 1983b) also indicates that there may have been at least three expansions of pastoralists into the East African highlands, the last two of which cannot, it seems, be correlated with any climatic shifts. However, our knowledge of the earlier parts of the pottery sequences is inadequate, and we cannot rule out the possibility of ceramic traditions diverging within the highlands (see Bower, chapter 6).

Conclusion

This chapter has examined briefly the way in which Africanist archaeologists often perceive the relationship between environment and culture and how environmental evidence has been used to explain cultural change. A tendency towards deterministic explanations rooted in the concept of "adaptation" has been noted. Two case studies have illustrated some of the problems involved in the marriage of archaeological and paleoenvironmental evidence; many of these problems stem from the paucity of data and sampling biases found in both disciplines. Clearly archaeologists must be cautious in their use of paleoenvironmental information, though this of course is a caveat that applies to all interdisciplinary endeavours. On a more positive note, archaeologists should strive towards a broader view of cultural change in which environment may be only one causative variable among many. As Salzman (1978:635) has indicated, "culture cannot construct reality, but it defines significance, and it is in terms of significance that men act." Similarly, Ingold (1983:14) has written that "the criteria of selection are not given by the environment alone but depend upon what members of the subject population are seeking to do in it. If the social domain is the source of their purpose, it follows logically that society *cannot* evolve through a process of adaptation." Therefore, social purposes and cultural concerns may steer a cultural system along a particular path among the many paths which environmental, ecological, demographic, historical, and other variables might make available. To approach problems of change from this broader and more complex perspective is the challenge that faces archaeologists.

References

Ambrose, S. H. 1982. Archaeology and linguistic reconstructions of history in East Africa. In *The Archaeological and Linguistic Reconstruction of African History* (eds C. Ehret and M. Posansky): pp. 104–57. Los Angeles: University of California Press.

_____. 1984a. The introduction of pastoral adaptations to the Highlands of East Africa. In *From Hunters to Farmers* (eds J. D. Clark and S. A. Brandt): pp. 212–39. Los Angeles: University of California Press.

_____. 1984b. *Holocene environments and human adaptations in the Central Rift Valley, Kenya*. (Unpublished Ph.D. thesis, University of California, Berkeley).

Binford, L. R. 1983. *In Pursuit of the Past: Decoding the Archaeological Record*. London: Thames and London.

Brandt, S. A. 1982. *A late Quaternary cultural/environmental sequence from Lake Besaka, Southern Afar, Ethiopia*. (Unpublished Ph.D. thesis, University of California, Berkeley).

Butzer, K. W., Isaac, G. L., Richardson, J. L. and Washbourn-Kamau, C. 1972. Radiocarbon dating of East African lake levels. *Science* 175:1069–76.

Clark, J. D. 1977. Prehistoric arrow forms in Africa as shown by surviving examples of the traditional arrows of the San Bushmen. *Paléorient* 3:127–50.

Clark, J. D. and Prince, G. R. 1978. Use-wear on Late Stone Age microliths from Laga Oda, Haraghi, Ethiopia, and possible functional interpretations. *Azania* 13:101–10.

Clarke, D. L. 1976. Mesolithic Europe: the economic basis. In *Problems in Economic and Social Archaeology* (eds G. Sieveking, I. H. Longworth and K. E. Wilson): pp. 449-81. London: Duckworth.

Coetzee, J. A. 1967. Pollen analytical studies in East and Southern Africa. *Palaeoecology of Africa* 3:1-146.

Collett, D. and Robertshaw, P. T. 1983a. Problems in the interpretation of radiocarbon dates: the Pastoral Neolithic of East Africa. *Afr. Archaeol. Rev.* 1:57–74.

_____. 1983b. Pottery traditions of early pastoral communities in Kenya. *Azania* 18:107-25.

Deacon. H. J. 1980. Late Pleistocene and Holocene industries in the southern Cape and wider correlations. In *Proceedings of the 8th Panafrican Congress of Prehistory and Quaternary Studies* (eds R. E. Leakey and B. A. Ogot): pp. 231–4. Nairobi.

Flannery, K. V. 1969. Origins and ecological effects of early domestication in Iran and the Near East. In *The Domestication and Exploitation of Plants and Animals* (eds P. J. Ucko and G. W. Dimbleby): pp. 73–100. London: Duckworth.

Gasse, F. 1980. Late Quaternary changes in lake-levels and diatom assemblages on the southeastern margin of the Sahara. *Palaeoecology of Africa* 12:333–50.

Godelier, M. 1972. *Rationality and Irrationality in Economics.* London: NLB.

Gramly, R. M. 1976. Upper Pleistocene archaeological occurrences at site GvJm/22, Lukenya Hill, Kenya. *Man* (n.s.) 11:319–44.

Hamilton, A. C. 1982. *Environmental History of East Africa: A Study of the Quaternary.* London: Academic Press.

Ingold, T. 1983. The architect and the bee: reflections on the work of animals and men. *Man* (n.s.) 18:1-20.

Inskeep, R. R. 1962. The age of the Kondoa rock paintings in the light of recent excavations at Kisese II rock shelter. In *Actes du 4e Congrès Panafricain de Prehistoire, Leopoldville, 1959* (eds G. Mortelmans and J. Nenquin). Tervuren.

Kamau, C. 1973. *Lake levels in the Rift Valley*. (Unpublished Seminar paper, Dept of History, University of Nairobi).

Lambrecht, F. L. 1964. Aspects of evolution and ecology of tsetse flies and trypanosomiasis in prehistoric African environment. *J. Afr. Hist.* 5:1–24.

Leakey, L. S. B. 1931. *The Stone Age Cultures of Kenya Colony*. Cambridge.

Leakey, M. D., Hay, R. L., Thurber, D. L., Protsch, R. and Berger, R. 1972. Stratigraphy, archaeology and age of the Ndutu and Naisiusiu beds, Olduvai Gorge, Tanzania. *World Archaeology* 3:328–41.

Mehlman, M. J. 1979. Mumba-Höhle revisited: the relevance of a forgotten excavation to some current issues in East African prehistory. *World Archaeology* 11:80–94.

Merrick, H. V. 1975. *Change in Late Pleistocene lithic industries in Eastern Africa*. (Ph.D. thesis, University of California, Berkeley). Ann Arbor: University Microfilms International.

Miller, S. F. 1979. Lukenya Hill, GvJm 46, excavation report. *Nyame Akuma* 14:31–4.

Osmaston, H. A. 1975. Models for the estimation of firnlines of present and Pleistocene glaciers. In *Processes in Physical and Human Geography. Bristol Essays* (eds R. F. Peel *et al.*): pp. 218–45. London: Heinemann.

Parkington, J. 1979. Review of D. W. Phillipson, The later prehistory of Eastern and Southern Africa. *S. Afr. Archaeol. Bull.* 34:73–74.

Phillipson, D. W. 1976. *The Prehistory of Eastern Zambia*. Nairobi: BIEA.

———. 1977. *The Later Prehistory of Eastern and Southern Africa*. London: Heinemann.

———. 1980a. Technological disparity, or the contemporaneity of diverse industries. In *Proceedings of the 8th Panafrican Congress* on *Prehistory and Quaternary Studies* (eds R. E. Leakey and B. A. Ogot): pp. 15-6. Nairobi.

———. 1980b. Some speculations on the beginnings of backed-microlith maufacture. In *Proceedings of the 18th Panafrican Congress on Prehistory and Quaternary Studies* (eds R. L. Leakey and B. A. Ogot): pp. 229-30. Nairobi.

Richardson, J. L. and Richardson, A. E. 1972. History of an African Rift lake and its climatic implications. *Ecological Monographs* 42:499–534.

Rudner, J. 1979. The use of stone artefacts and pottery among the Khoisan peoples in historic and protohistoric times. *S. Afr. Archaeol. Bull.* 34:3–17.

Salzman, P. C. 1978. Ideology and change in Middle Eastern tribal societies. *Man* (n.s.) 13:618–37.

Survey of Kenya. 1970. *National Atlas of Kenya*. Nairobi.

Van Noten, F. 1971. Excavations at Munyama cave. *Antiquity* 45:56–8.

———. 1982. *The Archaeology of Central Africa*. Graz: Akademische Druck.

Wayland, E. J. 1924. Palaeolithic types of implements in relation to the Pleistocene deposits of Uganda. *Man* 24:124.

Patterns of Environment Utilization by Late Prehistoric Cultures in the Southern Congo Basin

By Sheryl F. Miller, Pitzer College.

The southern Congo Basin is a geographical unit comprising the drainage of the Kwango and Kasai River systems, which flow mostly northward from their headwaters in Angola and Zaire, eventually sweeping west to join the Zaire (formerly Congo) River. Much of the area is a sloping plateau, with more hilly terrain to the south, in the Lunda region of Angola, and east of the Kasai River, where it separates Angola from Zaire. In general, the altitude is 500–1000 meters. The plateau is capped with sands and sandstones in some places as much as 200 meters thick (Haughton 1963). Where erosion has cut through this layer, the underlying metamorphic rocks of the Kasai Shield and the Dibaya Complex are revealed (Cahen and Snelling 1966). The present river pattern was established in the Upper Pliocene, with erosion occurring through most of the Quaternary. Much of the region is hot and humid, with temperatures a fairly constant 26° C near the equator. The average annual rainfall is 1200 to 1600 mm through much of the area, coming in two rainy seasons per year.

The soil chemistry in this environment is not conducive to the preservation of organic materials; thus archaeologists must rely strongly on the artifacts themselves and their geological context for clues to prehistoric human cultural-environmental relationships in the region.

At the time of the last glacial maximum there was an episode of significant aridity in the southern Congo Basin (Clark 1968). This may have been due to changing weather patterns, along with a northward shift of the cold Benguela Current in the Atlantic which creates a rain shadow across Angola and part of the Congo Basin. During this time sands from the Kalahari Desert were redistributed by wind northward into the Congo Basin, even reaching north of the river in Bas-Congo. This episode of sand deposition created environmental changes which outlasted the period of aridity itself. The initial environmental effect of this dry, cool period was a reduction in the extent of Congolese rain forest, and an expansion of savanna-woodland and open grassland areas. Even after the return of more moist conditions, the drainage system struggled to clear the blanket of sand. Where this redistributed sand and earlier sandstones still top the interfluvial plateau, they hinder the establishment of tree cover, as their porosity and thickness prevent tree roots from reaching water during the dry season. By contrast, the valley sides plunge steeply; where the underlying basement rocks are revealed there is a narrow band of deciduous woodland, but the main vegetation pattern of the valley bottom is a heavy forest growth. These two main zones—forested river valleys and interfluvial plateau grasslands—interfinger abruptly and form an environmental mosaic. In the more broken, hilly uplands this environmental mosaic is less striking, but there, too, a warm post-Pleistocene climate has encouraged a dense band of riverine forest with open woodland or grasses on the interfluves.

The modern indigenous inhabitants of the southern Congo Basin are primarily horticulturalists, but they make regular use of the fish and game in their environment, and supplement their diet with a variety of wild plant foods. They are aware of the resources available in both forests and grasslands, and extract them by different means. Hunting, in particular, most effectively utilizes different strategies. In the forest, a group of hunters may cooperate to drive the generally solitary, nocturnal game into nets where either the bow and arrow or short stabbing spear can be used. Rifles are now favored weapons for those who can acquire them, but older men recall the earlier hunting techniques. Although there are large animals such as the forest hog and even elephant available, the most common kill is probably the much less dangerous duiker. There are a number of forest duiker species, ranging up to 50 kg per adult animal. Occasionally an individual hunter among the BaYaka people of Zaire may go into the forest with his bow, shooting monkey or even bongo. In Lunda, during the dry season, individuals also seek prey in the open grassland where a BaChokwe hunter may hide by a game trail with his weapon poised.

These Congo Basin peoples use a wide variety of special purpose projectile points, nowadays made of iron. Transverse arrowheads are heavy and carry only a short distance, but leave a strong blood spoor that can be easily followed in the dense forest. In open grassland, on the other hand, arrows tipped with lighter foliate or tanged points carry farther with accuracy. In both terrains, the lanceolate-tipped spear may be used for throwing short distances, or for stabbing trapped game.

It is not difficult to believe that the environmental changes in the southern Congo Basin during the last glacial maximum and afterward might have had significant impact on human cultures. It may have been at this time that a Congolese lithic industry known as the Lupemban gave way to a complex now called the Tshitolian. A transitional "Lupembo-Tshitolian" proposed in the literature (Clark 1963) is included here as the earliest Tshitolian phase.

The Lupemban has been described (Clark 1963) as primarily a forest industry, comprising such tools as heavy core axes, picks, and core scrapers along with long, thick lanceolates. It is beyond the scope of this

However, there seems to have been a significant amount of cultural continuity between it and the subsequent Tshitolian. Therefore, it appears that the Tshitolian was a local cultural development, retaining many of the old ways but adding some new tools and technologies. A major technological development was the modification of an earlier prepared core into a core for blade production; these blades were retouched into a new, thinner style of lanceolate, and were also utilized in the manufacture of a new transverse projectile point: the *petit tranchet*. Other small projectile point types also appear in the Tshitolian, including a foliate form and another that is winged and tanged. These are customarily interpreted as arrowheads, implying that the bow was an innovation in local hunting technology which coincided with the advent of the Tshitolian. The Tshitolian also includes a new indirect punch technique as a means of primary blade production, as well as for the delicate, controlled retouching of such tools as flat lanceolates.

In the hilly upland country, Tshitolian sites are frequently associated with heavy sand deposits. Two site types are represented here: temporary camps where a small group stopped briefly to utilize special resources, and larger areas of dense artifact distribution which appear to indicate repeated, possibly long-term occupation. In northeast Angola, Lunda district provides numerous instances where a shallow lens of archaeological material occurs in otherwise sterile sands (Clark 1963). One such site, at Cauma Bridge, lies in a valley where a tributary stream joins the Chiumbe River. Excavations I conducted there in 1968, as part of a research program coordinated by J. D. Clark, indicate that Tshitolian toolmakers came to gather and work the *grés polymorphe* they utilized for nearly all their lithic implements. This raw material is found in the river gravels and was preferentially selected by both Lupemban and Tshitolian craftsmen. Experiments and interviews conducted in 1968 with Mwambumba, a local maker of gunflints for the flintlock rifles still in use, revealed that raw material which has remained damp works more reliably than the same *grés polymorphe* after it has dried out. Prehistoric toolmakers may well have noticed this characteristic of the stone, and if so would have come to use the cobble deposits along water courses such as the Cauma Bridge site. At Cauma Bridge the gravels themselves incorporate a range of artifacts from earlier prehistoric periods, as well as some apparently fresh Tshitolian materials. Other Tshitolian pieces occur scattered through sands immediately above the gravels. Their distribution suggests that the toolmakers may have sat on the bank of the sluggish, sand-choked stream to work. Sand deposits several meters thick subsequently covered and protected the site.

Also in the hilly Lunda area of northeast Angola are more intensively utilized sites. One is Mbalabala, near the town of Dundo, where a ridge overlooking the Luachimo River has a dense surface scatter of artifacts extending approximately a mile along the river. Excavations conducted here in 1968 by J. D. Clark yielded a horizon of early Tshitolian artifacts which had not yet been disturbed by erosion. This assemblage included such pieces as fine, thin lanceolates and other Tshitolian materials. This open hillslope site is primarily covered with coarse grass and a few scrubby trees. It has a good view of the river where hippos still are found, and of the narrow strip of riverine forest. These valley resources would be readily accessible to a band of hunter-gatherers living from time to time on the Mbalabala ridge. Clark (1963) has suggested that human-induced annual burning may have maintained the open character of such grassy slopes from Tshitolian times or even slightly earlier, and that prehistoric hunters might have exploited the available large biomass of elephant and hippo as well as smaller game.

Farther north, in the middle reaches of the Congo River tributaries where the geographical relief and environmental zones are more clearly defined, a sharper focus on the character and distribution of Tshitolian occurrences is possible. Sites here are of two basic varieties: those on the grassy plateau, and those down in the forested river valleys.

The Tshitolian type-site, on the Plateau of the Bena Tshitolo in the Kasai, represents an open grassland occurrence. In 1970 and 1971, I studied artifacts from this site and others housed at the Royal Central African Museum in Tervuren, Belgium. Characteristic artifacts from these plateau sites comprise small core axes, some flat lanceolates, and a range of foliate and tanged projectile points. *Petits tranchets* are absent from these assemblages. In the early 1970s, I was able to interview the Reverend V. A. Anderson about his discovery of the Tshitolian type-site in the 1930s (Breuil 1944); he recalled observing that similar sites were only found above a certain altitude, on the plateau. Such plateau sites appear rich in implements, and abundant debitage has been reported (M. Bequaert, field notes) although little exists in the museum collections studied. Therefore, it would seem that these sites were not simply single-function camps, but represent areas where a variety of activities were conducted by groups who may have returned again and again to the same place.

An equal intensity of occupation is suggested by the riverine forest Tshitolian occurrences. One such area is found at Ndinga Kiitu (formerly Ndinga St. Pierre) just above the confluence of the Sangunu with the Kwango River and well within the present valley forest zone. Maurice Bequaert (1952) excavated at Ndinga and other Kwango sites in the 1930s, and I studied his collections in 1970 and 1971 in Tervuren. At that time M. Bequaert generously shared with me his field notes and photographs of his research. Excavations conducted at Ndinga in 1973 by a Pitzer College team, invited by the National Museum of Zaire, verified Bequaert's findings. The Ndinga assemblage was rich in core axes, along with some other core tools. The lighter component of the tool kit included some lanceolate projectile points, but was dominated by the *petit tranchet*; no small foliate or tanged points were found in this forest Tshitolian site. There was a tremendous wealth of debitage in the Ndinga occurrence, and it is probable that toolmaking was one of the many functions undertaken in the valley, where *grés polymorphe* could be found. The abundance of core axes in a variety of shapes and sizes also suggests that woodworking was a major valley task where trees were readily accessible. Clark (1963) has conducted experimentation in Angola from which he inferred a woodworking function, in adze fashion, for the core axe. My own observations on edge-wear support that interpretation.

Thus the archaeological data speak for the presence at a large valley site such as Ndinga of several types of resource extraction: hunting with the bow and arrow tipped with the *petit tranchet*, the occasional utilization of a lanceolate-tipped spear, manufacture of wooden weapon parts and perhaps other wooden utensils as well, and the making of stone implements from the local *grés polymorphe*. The Tshitolian foragers may also have exploited resources used at present by the BaYaka, including such foods as wild fruits, roots, greens, mushrooms, and caterpillars, along with various fiber and medicinal products. The archaeological record, however, is mute about human use of such resources.

It should not be surprising that different resource utilization techniques, reflected in different tool kits, would have been employed in the different environments of the southern Congo Basin. Some questions of cultural dynamics nevertheless cannot be answered at this time. Were the same people utilizing both plateau grassland and riverine forest environments, perhaps seasonally, as indigenous people do today? Or were there two different, contemporaneous peoples in the two habitats as Bequaert and Mortelmans (1955) have suggested? Was there a basic chronological difference, with the heavier implements of a supposedly earlier valley occupation evolving into the smaller plateau forms? No multicomponent stratified site with both cultural facies in primary context has yet been excavated, and there are still too few radiocarbon dates from this region to establish a detailed cultural chronology. It does appear, however, that the Tshitolian succeeded the Lupemban at or shortly after the time of the last glacial maximum. The earliest dated Tshitolian material is 14,500 ± 560 bp, in northeast Angola (Clark 1968).

This fits with the geological evidence for the Tshitolian occurring in conjunction with the accumulation of redistributed Kalahari sands.

Were the cultural innovations observable in the Tshitolian an autochthonous development, possibly a local response to new environmental circumstances? Or were they introduced by movements of peoples and ideas, perhaps triggered by the shifting ecosystems?

Many elements of the Tshitolian first appeared in the southern Congo Basin coincident with the environmental change that brought increased grassland to the plateau areas. Blade technology and the indirect punch technique had been used earlier in other parts of Africa, from which the idea may have spread to the Congo Basin. This innovation enabled Tshitolian toolmakers to form the thinner, bifacial pieces which replaced heavy Lupemban lanceolate spear points.

New food procurement systems probably also came into the region with the advancing grassland. Most evident is the introduction of bow and arrow hunting, which was used earlier by the Nachikufan in open woodland just south of the Congo Basin (Miller 1971). However, Nachikufan microlithic technology and pointed backed bladelets were developed in an industry largely made on quartz, while Tshitolian toolmakers had available the flintlike *grés polymorphe*. The idea of a stone arrowhead may have entered their area with the new grassland hunting style, but Tshitolian toolmakers used their own raw material and cultural traditions when they created the bifacial foliate arrow points, like miniature lanceolates, found in the grassland facies.

The true invention of the Tshitolian was the *petit tranchet*. Adapting bow and arrow hunting to their forests, they retouched thin punched blades into this new arrowhead style. No evidence is presently known for the occurrence of the *petit tranchet* prior to the Tshitolian. Many specimens from Ndinga are irregular and varied in form, and thus appear to be early attempts; they contrast with other *petits tranchets* regularly made from blades with the ends snapped off and the edges backed.

Thus evidence suggests that the people who used Tshitolian lithics retained from the Lupemban a traditional forest-related tool kit of core axes and other heavy implements. They adopted new blade-processing technology and new bow-hunting techniques when climate change extended grassland into their territory. They adapted these to their own cultural terms, and made the significant invention of the *petit tranchet* which enabled them to use the bow effectively in forest hunting. The people of the Tshitolian thus had a dual resource base available in forest and grassland, along with an improved technology for resource extraction. They were probably successful hunter-gatherers in the region until the introduction of farming in the southern Congo Basin.

If these proposals meet the test of further investigation, more environmental information, and additional dates, then the advent and character of the Tshitolian will have been shown to result from a combination of environmental and cultural factors.

Acknowledgements

Thanks for making this research possible are due to many people and institutions. Foremost among these are Dr. J. Desmond Clark, who inspired the research, and my husband, Dr. Stephen A. Miller, who assisted in it. I am also grateful to my students and workers for their help in excavation, and to colleagues for their cooperation, especially Dr. Francis Van Noten and the late Maurice Bequaert in Belgium. For their financial and other support, gratitude is also expressed to the Ford Foundation, the Royal Central African Museum in Belgium, the National Museum of Zaire, and to Pitzer College.

References

Bequaert, M. 1952. Fouilles a Dinga (Congo belge). *Actes du Congres Panafricain de Prehistoire, Alger*: pp. 347–53.

———. and Mortelmans, G. 1955. Le Tshitolien dans le Bassin du Congo. *Academie Royale des Sciences Coloniales, Memoires in 8° Nouvelle serie 30* (2):1–39.

Breuil, l'Abbe H. 1944. Le Paleolithique au Congo Belge d'apres les recherches du Docteur Cabu. *Transactions of the Royal Society of South Africa* 30 (2):143–160.

Cahen, L. and Snelling, N. J. 1966. *The Geochronology of Equatorial Africa*. Amsterdam: North Holland Publishing Co.

Clark, J. D. 1963. Prehistoric cultures of northeast Angola and their significance in tropical Africa. *Diamang, Publicacoes Culturais* n° 62. Lisboa.

———. 1968. Further palaeo-anthropological studies in Northern Lunda. *Diamang, Publicacoes Culturais* n° 78. Lisboa.

Haughton, S. H. 1963. *The Stratigraphic History of Africa South of the Sahara*. London: Oliver and Boyd.

Miller, S. F. 1971. The age of Nachikufan Industries in Zambia. *South African Archaeological Bulletin* 26:1143–46.

———. 1972. A new look at the Tshitolian. *Africa-Tervuren* 13 (3/4):86–9.

Plate 8.1: Typical terrain of the southern Congo Basin. Evergreen primary forest in the valley is bordered by a narrow band of deciduous woodland, with open grassland on the interfluvial plateau.

Patterns of Environment Utilization by Late Prehistoric Cultures

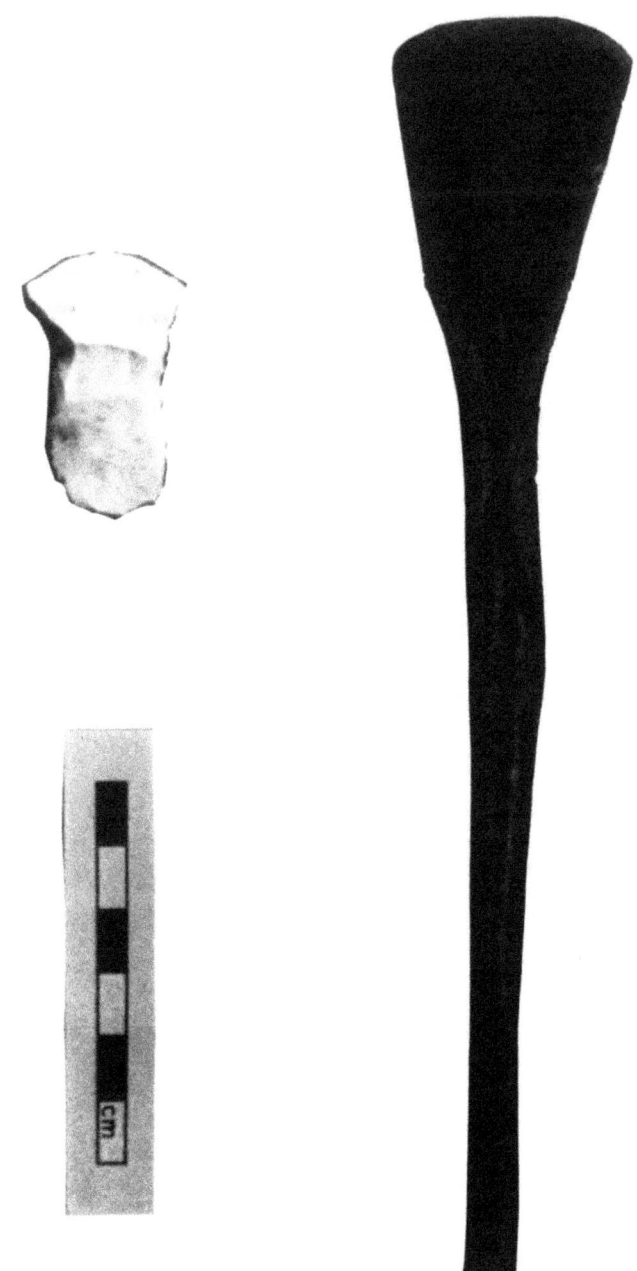

Plate 8.II: Transversal arrowheads. A Tshitolian *petit tranchet* from Ndinga Kiitu is on the left with a modern iron specimen from the same locale on the right. The BaYaka use weapons like the latter when hunting forest prey.

S. F. Miller

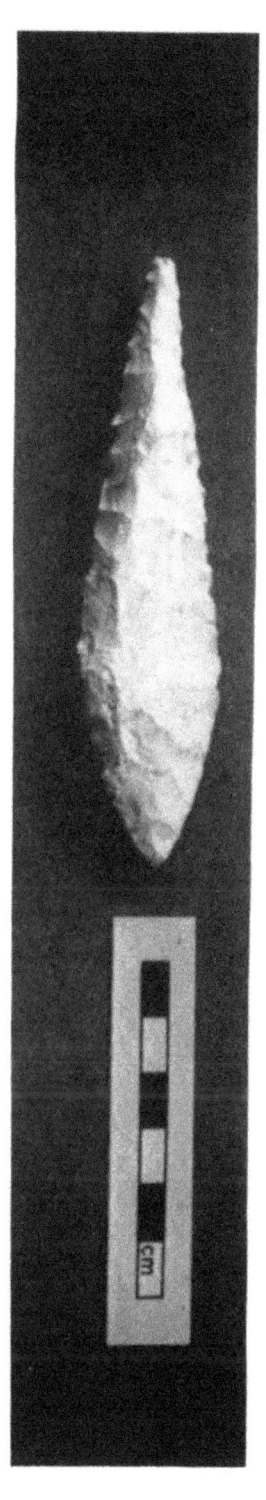

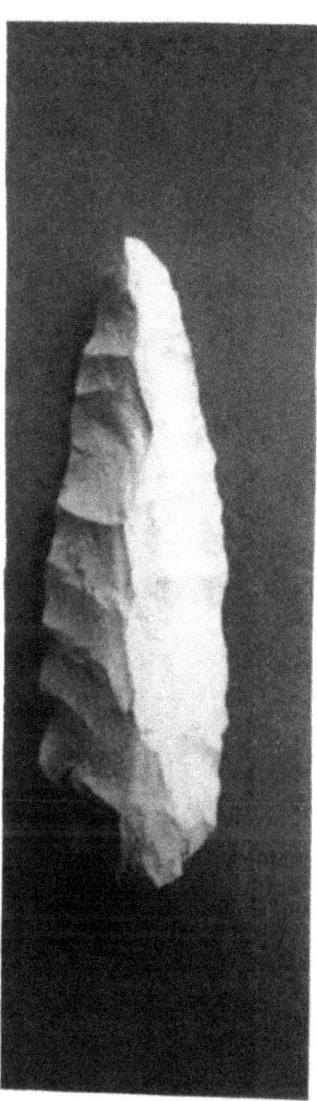

Plate 8.III: Foliate points. A modern BaYaka arrowhead is flanked by stone specimens from the Tshitolian. The BaYaka use this form largely for grassland hunting.

Patterns of Environment Utilization by Late Prehistoric Cultures

Plate 8.IV: A Tshitolian blade in *grès polymorphe*.

Plate 8.V: Tshitolian lanceolates. The denticulate specimen (third from left) is particularly thin, and was probably retouched by the indirect punch technique. The tips of the first and fourth specimens have been broken off.

Patterns of Environment Utilization by Late Prehistoric Cultures

Plate 8.VI: Excavations at Cauma Bridge. Each pit is two meters square. Many Tshitolian sites lie in deep sands deposited by winds in the late Pleistocene and subsequently reworked by water. The cut behind the pits shows this overburden. The piles of sand in the background are material which had been removed from the site by mining activities prior to archaeological excavation.

Plate 8.VII: Mwambumba, a Tchokwe maker of gunflints, demonstrating the indirect punch technique with which he replicates the blades and flake-scar patterns found in many Tshitolian assemblages.

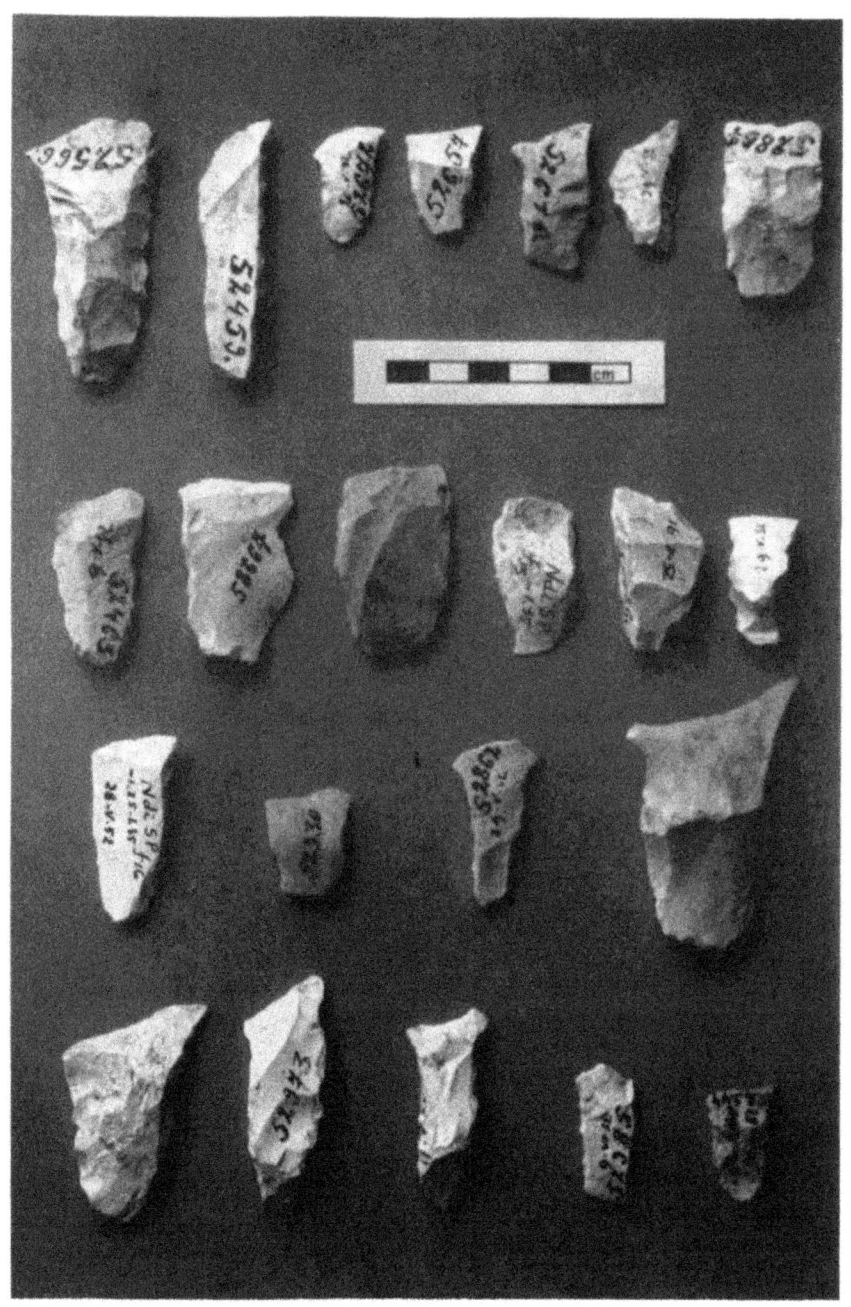

Plate 8.VIII: *Petits tranchets* from Ndinga Kiitu, illustrating the variations in size and form found at this forest site. These artifacts have in common their sharp transversal cutting edge and additional retouch, in the manner of backing, along the sides.

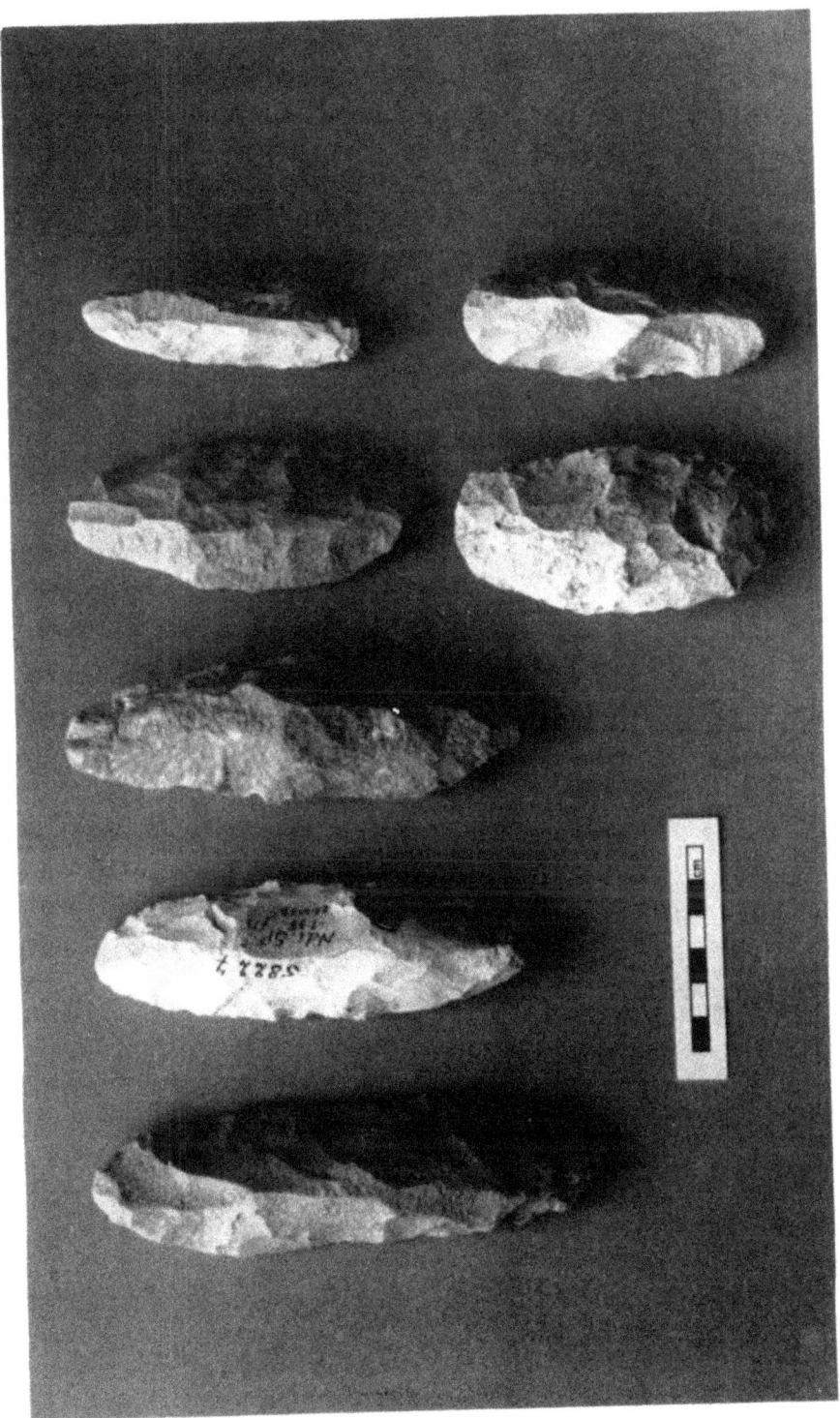

Plate 8.IX: Core axes from Ndinga Kiitu, showing a range of sizes and forms. The gouge-like working ends (pointing up, in the illustration) are often retouched, occasionally even after severe damage.

Patterns of Environment Utilization by Late Prehistoric Cultures

The Scale and Timing of Technological and Environmental Changes over the Last 20,000 Years in the Southern Cape, South Africa

By Janette Deacon, Department of Archaeology, University of Stellenbosch, South Africa.

Introduction

There has been no detailed investigation in southern Africa of the way people have responded culturally and technologically to the scale of environmental change that took place between glacial and interglacial conditions. Much of the discussion of the relationship between environment and technology has been synchronic, describing the contemporary distribution of formal tools and comparing this with the distribution of particular resources. Examples are the explanation of the distribution of Sangoan picks as a response to woodland environments (Clark 1959), the correlation of the distribution of Later Stone Age small convex scrapers and ethnographic records of leather clothing (Deacon and Deacon 1980) and the distribution of late Holocene adze-dominated assemblages that are assumed to relate to woodworking and the manufacture of digging sticks (Mazel and Parkington 1978).

Diachronic change has been noted in sequences dating to the last 20,000 years in the southern and eastern Cape Province in South Africa. It shows a relatively rapid shift in both technology and larger mammal remains at occupation sites dating to about 12,000 bp and a second series of technological changes between 8000 and 6000 bp (H. J. Deacon 1969, 1972; Klein 1972a, b). These were interpreted as responses to environmental changes that took place at the end of the Last Glacial Maximum (LGM) and at about the time of the mid-Holocene "climatic optimum" (H. J. Deacon 1972; Klein 1974). The reasoning was that if technological change is an adaptive device to cope with changes in the environment, then we should expect that if climatic shifts altered resources, new tools and technologies would be devised to adjust to the new circumstances. Another reason to expect change is that the kinds of environments in existence during the LGM have no modern analogue, and the food-getting strategies that were practised at that time by hunter-gathers would not have been possible in the later Holocene because plant and animal communities changed so significantly.

To test in detail the relationship between technological and environmental change, two well-dated, long-sequence Later Stone Age sites in the southern Cape Province, Nelson Bay Cave on the coast and Boomplaas Cave 80 km inland, were selected for study. Essential to the success of the project was the association of a number of different environmental indicators, both biological and geological, and the

Scale and Timing of Technological and Environmental Changes

occurrence of similar stone artefact sequences at both sites. Prior to the excavation of these sites, the quality of environmental data available in the 1960s and early 1970s was not good enough to test the relationship between environmental and technological change adequately, although various models were suggested (H. J. Deacon 1976; J. Deacon 1978). An excavation programme at Boomplaas Cave was begun in 1974 specifically to investigate the relationship (H. J. Deacon 1979). The research has shown that the scale and timing of environmental changes have been complex and that technological changes reflect as strong a stylistic as a functional component.

It is important to test the environmental model because the time scale involved precludes the use of ethnographic analogies for such adaptive behaviour. Where historical data on climatic and technological change have been analysed in the Middle East, no causal relationship could be found (Bowden *et al.* 1981:489) and the interacting variables were so complex that the exact importance of particular factors involved in demographic and technological change could not be determined. Later Stone Age people undoubtedly lived a less complex life than their Middle East counterparts 4000 years ago, but the environmental model is at best simplistic and at worst untrue and needs to be rigorously examined before it can be retained as a viable explanation for technological change.

The period selected for comparing the scale and timing of changes at the two sites is the last 20,000 years, covering the Last Glacial Maximum (LGM) when temperatures were about 5–6° C cooler than at present, the terminal Pleistocene when temperatures became warmer and rainfall increased, and the Holocene which saw the warmest conditions of the last 120,000 years. The cultural record includes the beginning of the Later Stone Age, the development of this technological tradition during the Holocene, and the introduction of domesticated stock within the last 2000 years. Thus, with significant and measurable changes taking place in both the environmental and cultural/technological spheres, we can expect that if there were indeed adaptive responses to environmental changes, these should be identifiable in the excavated materials from the two sites and the timing of these changes should coincide in the two sequences.

The Environments

Both Boomplaas Cave (BPA) and Nelson Bay Cave (NBC) are situated in the southern Cape, a temperate region with all-year rainfall and peaks in spring and autumn, but whereas the annual rainfall in the vicinity of NBC is between 600 and 700 mm, it is only 400 mm at BPA.

Nelson Bay Cave is a large cave on a rocky shore and was excavated in the 1960s and 1970s by Inskeep (1966, 1972) and Klein (1972a; J. Deacon

1984a). A sequence of deposits dating from the Middle Stone Age (MSA) through the Later Stone Age (LSA) was found. The data used here are from the post-MSA levels in Klein's excavations and date to between 18,000 and 5000 bp. Environmental information has been drawn from the sediments (Butzer 1973, 1984), from the large mammals hunted by people occupying the cave (Klein 1972a, b), from micromammals eaten by owls (Avery 1982), and from marine shells collected for food by the LSA inhabitants (Klein 1972a; Shackleton 1973).

The larger mammals indicate that between 18,000–12,000 bp grassland vegetation predominated in the vicinity of the cave in contrast to the coastal fynbos around the site at present. Grassland was encouraged both by the lower temperatures at the Last Glacial Maximum (LGM) and by the exposure of a flat coastal plain south of NBC, where lower sea levels associated with the LGM moved the coastline some 80 km south of its present position. The fauna includes large gregarious grazers such as the extinct giant buffalo, *Pelorovis antiquus*, the extinct giant Cape horse, *Equus capensis*, an extinct giant hartebeest, *Megalotragus priscus*, as well as springbok and other extant grassland species. Micromammalian remains in the deposits dated to between 18,000 and 16,000 bp allow a more detailed reconstruction suggesting that the LGM in the area was not only cold, but also dry, although Butzer (1973, 1984) has concluded from the sedimentological analysis that the environment around the cave was wet rather than dry. This is contrary to other evidence from both Nelson Bay and Boomplaas.

By 12,000 bp the sea level had risen sufficiently to make it worthwhile to exploit shellfish from the site, although the species composition of the shell midden shows that sea temperatures were still cooler than at present. Dated littoral shells dredged offshore from Cape St. Francis to the east of Nelson Bay give some idea of the speed and dating of the rising sea level along the southern Cape coast (Dingle and Rogers 1972) (Fig. 9.1). The larger mammals show the transition from grassland to bushier vegetation between 12,000 and 10,000 bp and by 8000 bp the assemblage is dominated by small nongregarious browsers such as bushbuck and grysbok, with significant numbers of seal, and large quantities of fish and shellfish (Klein 1972a). The same species carry through to the end of the Holocene with only minor variations in relative frequencies.

Boomplaas Cave is situated in a band of limestone in a small valley with a permanent stream on the south-facing slopes of the Swartberg range. The limestone has helped to preserve bone in the deposit and the rapid accumulation of sterile units has been useful in separating those built up by human and nonhuman occupation. Excavations from 1974–1979 were designed to recover a range of different environmental indicators (H. J. Deacon 1979; Deacon *et al.* 1984). The sequence dates within the last 80,000 years, but the deposits discussed here date from 21,000 bp to 1500 bp. Analyses have been completed on the large mammal fauna (Klein 1978; Von den Driesch and Deacon 1985), the micromammalian remains (Avery 1982), charcoal left in hearths (Deacon *et al.* 1983), pollen blown into the cave (Scholtz in prep.), the cave sediments (Webley 1978), and the oxygen isotope ratio sequence in a speleothem from the nearby Cango Caves (Vogel 1983).

As at NBC, the deposits dating between 21,000 and 12,000 have large mammal remains characteristic of a grassland vegetation including the extinct species also present at NBC. The relative abundance of browsers is therefore low (Fig. 9. 1). The micromammalian remains have been particularly informative at BPA because they were recovered in significant quantities from most of the stratigraphic units. Analysis of the community composition of samples in this time range confirms the conclusions drawn from the larger mammals, namely that conditions at the Last Glacial Maximum (25,000–16,000 bp) were cold and dry, but show also that a major change to less harsh conditions took place between 16,000 and 14,000 bp (Avery 1982), at least 2000 years earlier than apparent in the larger mammal record. The timing of change in the community composition of micromammals is similar to changes in the size of individuals of the red musk shrew, *Crocidura flavescens*, which are significantly larger at the LGM than under warmer conditions (Avery 1982). Factor analysis of the frequencies of micromammalian taxa in each excavated unit has grouped certain species that are relatively common in Holocene deposits. These taxa, having high loadings on Factor 1, contrast with another group of species which have low loadings on the same factor and which are today distributed in relatively cooler regions of the subcontinent. Values from 1 to 100 have been calculated from the factor scores to draw the histogram in Fig. 9. 1. It is not an absolute measure of temperature change, but gives an idea of the scale and timing of change linked to temperature and other environmental features that characterized the Holocene as opposed to the Late Pleistocene. The lowest diversity in the micromammals is at 18,000 indicating the harshest conditions at the LGM. Factor analysis of the Boomplaas microfauna indicates that this period was also relatively dry (Deacon *et al.* 1984).

Further confirmation that conditions at the last Glacial Maximum were cold and dry at Boomplaas is shown in the species composition of charcoal samples from hearths dated between 21,000 and 18,000 bp. The range of taxa is small and there are few trees represented in samples that are dominated by Compositae including the small shrub renosterbos, *Elytropappus rhinocerotis* (Fig. 9. 1). Detailed analysis of the anatomy of the charcoaled wood by Scholtz (in prep.) further indicates the environmental stress of plant communities of the LGM relative to the Holocene. The dominance of *Elytropappus* in the vegetation surrounding the site is mirrored in the pollen from the same levels (Deacon et al. 1983). The charcoal and micromammal data are in complete agreement, and there is marked amelioration of conditions bracketed between 16,000 and 14,000 bp that includes not only warmer temperatures, but also higher rainfall in the Cango Valley.

Roof spalls are more common in the LP Member dating to 21,000 bp than they are in other stratigraphic units and although clast-supported clastic rubble or true *éboulis sec* was not forming (possibly because winter drought reduced the chance of frost weathering), there was a greater incidence of spalling at this time than at any other. Cold temperatures are also indicated by the oxygen isotope ratio values in a speleothem from the nearby Cango Caves where the samples dated to between 20,000 and 18,000 bp indicate the coldest temperatures of the past 35,000 years estimated at 5° C cooler than at present (Vogel 1983) (Fig. 9. 1).

As in the microfauna, changes took place in the charcoal, pollen, and cave sediments between about 18,000 and 14,000 bp and by 14,000 bp indications are that conditions in the vicinity were much wetter and somewhat warmer. Woody taxa such as *Rhus* sp. and *Olea* sp. dominate the charcoal samples (Fig. 9. 1) suggesting even more effective precipitation than at present. The oxygen isotope values in the speleothem from Cango show a rise in temperature at about 15,000 bp and thereafter the speleothem stopped growing, as did a stalagmite in Boomplaas Cave that was active from the base of the deposit at about 80,000 bp to the base of the CL unit dated to 14,200 bp. The speleothems started growing again in the mid-Holocene. The explanation may lie in the fact that the speleothem record is only in part temperature related and may reflect changes in rainfall patterns. In this sense it may be a synoptic climate indicator. The rise in temperature before 14,200 bp is in tune with evidence in other southern hemisphere continents for a general post-glacial maximum warming which preceded that in the northern hemisphere by several thousand years (Salinger 1981).

The larger mammal fauna show no appreciable change between 18,000 and 14,200 bp, but there is a marked shift from grazers to browsers between 12,000 and 10,000 bp that is similar to the one observed at NBC (Fig. 9. 1). Thereafter the Holocene samples are dominated by small nongregarious browsers and, after about 1800 bp, by domesticated sheep. There are some changes in species composition of charcoal and micromammal samples during the Holocene, but these are not at the same scale as those observed between the LGM and the post-Glacial. The charcoal shows drier conditions at both 10,000 and 6400 bp. The

Scale and Timing of Technological and Environmental Changes

latter is the only mid-Holocene occupation unit and both the charcoal and the microfauna indicate the warmest conditions of the whole sequence. Slightly cooler temperatures prevailed in the overlying units dating within the last 2000 years (Deacon *et al.* 1984). It is only in these post-6500 bp deposits that modern plant and animal communities are reflected in the floral and faunal remains.

The paleoenvironmental record from both sites therefore confirms that the LGM was cold and dry. Thereafter, the Cango speleothem, the micromammals, the charcoal, and the pollen records show an amelioration of temperature between 16,000 and 14,200 bp that was accompanied, on the evidence from BPA, by much wetter conditions than those of today. Always assuming that the relative frequency of browsers and grazers in the larger mammal fauna reflects their relative abundance in the vicinity of the sites, the larger mammals did not respond to these temperature and rainfall changes until after 12,000 bp when several giant species became extinct. Klein (1984) sees these extinctions as the result of a combination of environmental stress and improved hunting technology, and indeed the late response of the larger mammals is more a reflection of human selection than of their abundance in the landscape. After 12,000 bp rainfall decreased and temperatures continued to rise to a peak in the early mid-Holocene. Late Holocene temperatures and rainfall ranges may have been similar to the present but synoptic climates were not necessarily modern.

There is not the same continuity of observations and complementary lines of evidence from any other sites in the country and these southern Cape data serve as a model against which to test observations elsewhere. Insofar as it has been tested (Tusenius 1986), the moister end-Pleistocene, drier early Holocene, and synthesis of modern plant and animal communities in the last 5000 years is confirmed from East Griqualand in the summer rainfall region about 500 km to the northeast. We should therefore expect to see technological change at periods of greatest environmental change, i.e., between 16,000 and 14,000 bp, possibly with the demise of the giant bovids between 12,000 and 10,000 bp, and possibly after the mid-Holocene when plant and animal communities took on a modern character.

The Artefacts

The stone artefact analysis was designed specifically to highlight changes that took place through time in the habits and traditions of stone tool manufacture. Aspects analysed included the choice of raw material, the flaking method used, the pieces selected for utilization, the range of formal tools made and their characteristics, and the pattern of discard. Assemblage characteristics were measured both by relative frequencies of stone artefacts in various classes and categories, and by metric attributes of untrimmed flakes among the informal tools, and of scrapers and adzes among the formal tools (for details see J. Deacon 1984a).

The range of tasks for which the stone tools were designed did not change radically over the period of time concerned. The toolkit was for the most part unspecialized and most stone tools are presumed to have been made for the manufacture and maintenance of items made from other raw materials such as wood, skin, and bone. The typology in general use in southern Africa places greater emphasis on formal tools than it does on untrimmed and utilized pieces because the formal component represents the most easily recognizable patterned behaviour. In fact, the formal stone tools, with a few exceptions, were not used directly in the food quest and are therefore probably poorly correlated with environmental change. Those that were linked to hunting and gathering, such as bored stone digging stick weights, upper and lower grindstones, and parts of bows and arrows, are relatively rare in rock shelter deposits and have not been measured for change through time. More detailed information on the use of stone tools for plant and animal foods could be obtained if artefact studies included both microwear and residue analyses. Microwear analysis of backed tools was limited, but we assume they were used as inserts in cutting tools and as inserts for spears or arrows, in which case they are the artefacts most closely linked to the food quest.

Comparison of microwear on other LSA artefacts and replicas (Binneman 1984) has confirmed that scrapers, the most important formal stone tool numerically, were employed primarily for working skins used for clothing and bags. As such, scrapers are related to dress styles (Deacon and Deacon 1980) and the size and style of manufacture may well have symbolic social overtones. Adzes were used for working wood for digging sticks, bows, and parts of traps.

Change has taken place at two levels: innovations in which new items appear in the archaeological record, and post-innovation changes in which modifications occur in the frequency, size, and shape of artefacts after they have been introduced. Using a purely functional interpretation, we would expect that innovations would represent new solutions to old tasks, or new tasks altogether, whereas post-innovation changes would represent smaller scale adjustments to the original model either to improve efficiency or to adjust to changing circumstances. In modern technological change, Elster (1983:93) makes a similar distinction between innovation and substitution, where substitution is a change in the production process using existing technical knowledge.

Scale and Timing of Technological and Environmental Changes

When the sequences from NBC and BPA and other sites such as Kangkara, Wilton, Melkhoutboom, Oakhurst, and Matjes River are compared, most of the major changes in relative frequency and metric attributes (where known) occur within less than a millennium at all sites, but individual attributes, including both innovations and post-innovation changes, alter at different rates through time. These adjustments occurred on a regional scale in the southern Cape and may be more widespread. They are not site specific so the stimuli for change must be sought on a regional or broader scale. The fact that individual attributes do not always change in tandem would suggest that a number of different stimuli initiated the changes.

The shift from MSA to LSA toolmaking traditions is essentially a change characterized by the appearance of a range of innovations. The new items do not all appear at once, nor are they necessarily all found in every assemblage, but they were all items that were still being made by LSA San and Khoikhoin people at the time of European contact. They therefore represent the tools of what Goodwin and Van Riet Lowe (1929) referred to as "Neoanthropic Man." The innovations include decorative items such as ostrich eggshell beads, shell pendants and beads, painted and engraved stones, tortoiseshell bowls, ostrich eggshell water containers (some decorated with incised lines), polished bone points, bored stones, grooved stones, and hafted microlithic stone tools. The criterion for identifying change at this macro-level of innovation is the dating of the initial presence of these items in long sequence deposits. Once they have appeared in the sequence, subsequent changes in frequency or design may occur but these are not considered to be part of the phenomenon of innovation. A good example is the introduction of worked ostrich eggshell in the form of beads, pendants, and decorated fragments of water flasks. Data for BPA are given in Fig. 9.1 using the number of pieces per bucket of deposit excavated. The first pieces occur in units dated to 14,200 bp and thereafter the frequency fluctuates. The timing of the innovation is assumed to have been between 17,000 and 14,000 bp because a few beads were found at NBC dating to between 18,000 and 16,000 bp. Another example is the introduction of pottery in the last 2000 years. In this instance the speed and direction of the innovation can be traced. Radiocarbon dates for the first appearance of pottery throughout Africa show that it filtered slowly from North Africa to the equator over at least four millennia, but thereafter appeared virtually simultaneously from the Congo to the southernmost tip of the continent within a few hundred years (Deacon 1984a, b). This type of rapid dissemination has also been noted in modern technology and is characterized by people moving with the innovation (Hodgen 1974).

The high incidence of formal tools may be an innovation of a different kind—possibly in the method of hafting rather than in the components of the hafted tools themselves. In this instance the radiocarbon dates for the earliest appearance of microliths in quantity in Holocene assemblages show a much more gradual spread. If the dates can be taken at face value, it took nearly 4000 years for the typical Wilton-type assemblage to spread from Zimbabwe to the southern Cape coast (Deacon 1984a). The mechanism for the introduction of this innovation is more likely to have been by gradual diffusion along trade and communication networks than by people themselves moving southwards. At BPA and NBC there is a marked increase in the incidence of formal tools between 7000 and 6000 bp (Fig. 9. 1). The fact that high formal tool frequencies start at different times in different areas in South Africa makes it difficult to believe that the introduction was triggered by climatic or environmental factors. High formal tool frequencies correlate with a diminution in the size of untrimmed flakes and with a slight increase in the number of bladelet cores at BPA and NBC (Fig. 9. 1), but are independent of raw material changes. No correlation could be found between the incidence of formal tools and the incidence of particular raw materials (Deacon 1984a). This particular phenomenon would therefore be the result of diffusion and borrowing.

The diminution in the size of stone artefacts through time is a feature common to many technological systems, including modern ones. Van Wyk (1979) has identified four evolutionary trends in modern technological systems: increasing complexity, increasing efficiency, improved size characteristics, and improved time characteristics. These can be identified in LSA technology in the hafting of formal tools (greater complexity), in the standardization of artefact design (increased efficiency) and in the trend towards smaller artefacts (improved size characteristics). Improved time characteristics are more difficult to demonstrate in the stone age record and must be assumed. The stone age data cover a much longer time span than those of modern technology, however, and show some interesting contrasts. Although there is indeed a long-term trend towards smaller tools, change is not regular. The increase in artefact size between 12,000 and 8000 bp and again at some places within the last 2000 years is counter to the long-term trend suggesting that factors other than evolutionary ones are involved. The most likely explanation is stylistic variation that is reflected in cyclic shifts in the length of flakes throughout the stone age (Beaumont 1978; J. Deacon 1984a).

Flaking technique is an aspect of stone tool manufacture that we would expect to see changing at the post-innovation level in terms of Elster's definition of "substitution." Changes in the frequency of bladelets at Boomplaas and Nelson Bay are illustrated in Fig. 9. 1 and show that the heyday of bladelet manufacture was from about 18,000 to about 12,000 bp, although they were made both before and after this time. What distinguishes them from those made at other times is their exceptionally high frequency, not simply their presence (J. Deacon 1984a). Whereas their first appearance in the LSA sequence has no apparent environmental correlate, their sudden and almost total disappearance at about 12,000 bp may well be linked to the change in hunting practice with the extinction of the giant bovids and the change to smaller browsing antelope. What it may reflect is a change in hunting weapons at this time.

Changes in the production process can also be seen in the mean length of untrimmed flakes, again illustrated for both BPA and NBC in Fig. 9. 1. This increases markedly between 12,000 and about 8000 bp at both sites and is reflected too in changes in the mean length of scrapers over the same time. It would seem that flakes were made larger deliberately and suitable raw materials were selected for this purpose. In a linear regression test between the length of untrimmed flakes and the percentage of quartzite in a sample, $P = .01$ in the NBC samples and $P = .01-.001$ in the BPA samples, the longer flakes correlating with higher percentages of quartzite (J. Deacon 1984a:113). The diminution in flake length from the mid-Holocene is similarly correlated with raw materials, this time with those that occur in smaller nodules such as quartz and chalcedonies.

Summary and Conclusions

If technological change was triggered by environmental shifts, then a major change should have occurred between 16,000 and 14,000 bp and more minor adjustments should be discernible around the mid-Holocene climatic optimum. Worked ostrich eggshell, an example of an innovation, appears at about 16,000 bp at NBC and at 14,200 bp at BPA, and the increase in formal tools just before the mid-Holocene could also be significant. On the other hand, pottery, another innovation, appeared independently of marked environmental change about 1800 bp. Seen in a broader subcontinental perspective, the fact that the increase in microlithic formal tools occurs nearly 4000 years earlier in Zimbabwe and nearly 2000 years earlier in the northern Cape than it does in the southern Cape (J. Deacon 1984a, b) makes the apparent environmental correlation in the southern Cape less convincing. The appearance of the Wilton technology in that region during the mid-Holocene was rather the end of a long route of diffusion along established social networks and its correlation with mid-Holocene warming is largely coincidental.

In the case of post-innovative or substitution changes, bladelets increase in frequency between 18,000 and 12,000, and untrimmed flake and scraper lengths increase between 12,000 and 8000 bp. These changes are part of the same continuum from smaller to larger flakes in the terminal Pleistocene and a subsequent return to smaller ones in the Holocene. Although it may be argued that the increase in scraper and untrimmed flake length and in the associated raw material changes at 12,000 bp signal a delayed response in the artefacts that correlates with the demise of the giant bovids, the return to smaller artefacts after 8000 bp has no similar correlate; neither does the initial onset of bladelet manufacture around 20,000 bp.

The change in the lengths of flakes clearly post-dates by at least 2000 years temperature, rainfall, and vegetation shifts associated with the end of the Last Glacial Maximum. The lag may signify that changes in hunting and gathering practices, and the associated social adjustments, could have been precipitated by earlier environmental shifts (H. J. Deacon 1976; J. Deacon 1978) but if environment was indeed a prime mover then it worked in a slow and devious way. Technology does not change only as a result of environmental stimuli, and here the mid and late Holocene adjustments are cases in point. The link between climatic change and changes in stone age technology shows that environment is not an initiator of technological change, but is rather a constraint on the distribution and densities of human populations (J. Deacon 1974; Deacon and Thackeray 1984). How they respond technologically is determined by widespread cultural norms. The distribution of industries reflects whether the range of strategies associated with the artefact tradition were appropriate to the environment at a particular time or place. If the technology was inappropriate for a particular place, people did not live there. When the environmental constraints changed or innovations were introduced to the artefact system, then distributions of people could change accordingly.

The southern Cape sequences show that there is little evidence to support any hypothesis that environment determines technology or that environmental change stimulates technological change in any direct way. The relationship is complex and there is no consistent connection between the two. The direction that technological change takes has as much to do with the options open at any particular time as with the inventiveness and receptiveness of the toolmakers. From the study of modern technological change, the more rapid the assimilation of a new item, the more likely it is that it was introduced by itinerant people. Slower changes are more likely to have been passed on gradually by diffusion and borrowing (Hodgen 1974). Whereas the emphasis has been on tracing influences that may have changed technology from outside the system, it is time now to explore the interrelationships of stone age social systems and technology, and the factors that are common to prehistoric and modern technology.

Scale and Timing of Technological and Environmental Changes

The problem arises when one tries to put such an intellectual recommendation into operation. Southern African stone age archaeologists have been castigated for not drawing hypotheses from the ethnography to interpret stone artefact assemblages in terms of social and ritual behaviour (Lewis-Williams 1984, unpubl.). The modern ethnography has been invaluable in interpreting the rock art and aspects of the diet and demography of LSA people, but there certainly has been a gulf in understanding between modern and prehistoric San technology that deserves to be narrowed. Unfortunately, the San interviewed by twentieth-century ethnographers have not made stone tools for at least 100 years. Stone toolmakers who were interviewed in the nineteenth century gave very little direct information and any hypotheses we may set up will of necessity be based on conjecture, particularly as regards diachronic change. Synchronic aspects may be more easily interpreted because there are references to inter-group differences in artefact manufacture made by the San informants of Bleek and Lloyd (1911, unpubl.) that could be investigated in the field.

The factors that are common to modern and prehistoric technological systems are perhaps more easily traced, but are problematic to interpret. It is difficult to accept the concept that technological systems have a momentum of their own and many object to the use of models that draw parallels between biological and technological evolution. This is because neither concept is capable of explaining the scale and timing of changes; they merely describe them. Even analysts of modern technological change, where there is far greater potential for tracing cause and effect, have difficulty in explaining fully the reasons why some innovations succeed and others do not (Elster 1983). Theories that draw on historical methodology may be more appropriate for the study of diachronic change in stone age technological systems than the testing of cause and effect using hypotheses drawn from an ethnographic record that contains only a few tantalizing clues.

Acknowledgements

I am grateful to H. J. Deacon for discussions and comments on drafts of this paper. Research at Boomplaas Cave was funded by the Human Sciences Research Council, Pretoria, in grants awarded to H. J. Deacon from 1974 to 1979. Research at Nelson Bay Cave was funded by NSF Grant GS-3013 awarded to R. G. Klein. I am grateful for their financial support.

References

Avery, D. M. 1982. Micromammals as palaeoenvironmental indicators and an interpretation of the late Quaternary in the southern Cape Province, South Africa. *Annals of the South African Museum* 85:183–374.

Beaumont, P. B. 1978. *Border Cave.* (Unpublished M.A. thesis, University of Cape Town).

Binneman, J. 1984. Mapping and interpreting wear traces on stone implements: a case study from Boomplaas Cave. In *Frontiers: Southern African Archaeology Today* (eds M. J. Hall, G. Avery, D. M. Avery, M. L.Wilson and A. J. B. Humphreys): pp. 143–151. Oxford: BAR International Series 207.

Bowden, M. J., Kates, R. W., Kay, P. A., Riesbank, W. E., Warrick, R. A., Johnson, D. L., Gould, H. A. and Weiner, D. 1981. The effect of climate fluctuations on human populations: two hypotheses. In *Climate and History: Studies in Past Climates and their Impact on Man* (eds T. M. L. Wigley, M. J. Ingram, and G. Farmer): pp. 479–513. Cambridge: Cambridge University Press.

Butzer, K. W. 1973. Geology of Nelson Bay Cave, Robberg, South Africa. *South African Archaeological Bulletin* 28:97–110.

Butzer, K. W. 1984. Late Quaternary environment in South Africa. In *Late Cainozoic Palaeoclimates of the Southern Hemisphere* (ed J. C. Vogel): pp. 235–264. Rotterdam: Balkema.

Clark, J. D. 1959. *The Prehistory of Southern Africa.* Harmondsworth: Penguin.

Deacon, H. J. 1969. Melkhoutboom Cave, Alexandria District, Cape Province: a report on the 1967 investigation. *Annals of the Cape Provincial Museums* 6:141–169.

_____. 1972. A review of the post-Pleistocene in South Africa. *South African Archaeological Society Goodwin Series* 1:26–45.

_____. 1976. *Where Hunters Gathered: A Study of Holocene Stone Age People in the Eastern Cape.* Claremont: South African Archaeological Society Monograph 1.

_____. 1979. Excavations at Boomplaas Cave—a sequence through the Upper Pleistocene and Holocene in South Africa. *World Archaeology* 10:241–257.

_____. and Deacon, J. 1980. The hafting, function and distribution of small convex scrapers with an example from Boomplaas Cave. *South African Archaeological Bulletin* 35:31–37.

_____., Deacon, J., Scholtz, A., Thackeray, J. F., Brink, J. S. and Vogel, J. C. 1984. Correlation of palaeoenvironmental data from the Late Pleistocene and Holocene deposits at Boomplaas Cave, South Africa. In *Late Cainozoic Palaeoclimates of the Southern Hemisphere* (ed J. C. Vogel): pp. 339–352. Rotterdam: Balkema.

_____. and Thackeray, J. F. 1984. Late Pleistocene environmental changes and implications for the archaeological record in southern Africa. In *Late Cainozoic Palaeoclimates of the Southern Hemisphere* (ed J. C. Vogel). Rotterdam: Balkema.

_____., Scholtz, A. and Daitz, L. D. 1983. Fossil charcoal as a source of palaeoecological information in the fynbos region. In *Fynbos Palaeoecology: A Preliminary Synthesis* (eds H. J. Deacon, Q. B. Hendey and J. J. N. Lambrechts): pp. 174–82. Pretoria: National Scientific Programmes Report 75.

Deacon, J. 1974. Patterning in the radiocarbon dates for the Wilton/Smithfield complex in southern Africa. *South African Archaeological Bulletin* 29:3–18.

_____. 1978. Changing patterns in the late Pleistocene/early Holocene prehistory of southern Africa as seen from the Nelson Bay Cave stone artifact sequence. *Quaternary Research* (NY) 10:84–111.

_____. 1984a. *The Later Stone Age of Southernmost Africa*. Oxford: BAR International Series 213.

_____. 1984b. Later Stone Age people and their descendants in southern Africa. In *Southern African Prehistory and Paleoenvironments* (ed R. G. Klein): pp. 221–328. Rotterdam: Balkema.

Dingle, R. V. and Rogers, J. 1972. Pleistocene palaeogeography of the Agulhas Bank. *Transactions of the Royal Society of South Africa* 40:155–165.

Elster, J. 1983. *Explaining Technical Change*. Cambridge: Cambridge University Press.

Goodwin, A. J. H. and Van Riet Lowe, C. 1929. The Stone Age cultures of South Africa. *Annals of the South African Museum* 27:1–289.

Hodgen, M. T. 1974. Anthropology, history and cultural change. *Viking Fund Publications in Anthropology* 52:1–108.

Inskeep, R. R. 1966. University of Cape Town excavations at Plettenberg Bay. *Scientific South Africa* 2:575–577.

_____. 1972. Nelson's Bay Cave, Robberg Peninsula, Plettenberg Bay. *Palaeoecology of Africa* 6:247-248.

Klein, R. G. 1972a. Preliminary report on the July through September 1970 excavations at Nelson Bay Cave, Plettenberg Bay (Cape Province, South Africa). *Palaeoecology of Africa* 6:177-208.

_____. 1972b. The late Quaternary mammalian fauna of Nelson Bay Cave (Cape Province, South Africa): its implications for megafaunal extinctions and environmental and cultural change. *Quaternary Research* (NY) 2:135-42.

_____ 1974. Environment and subsistence of prehistoric man in the southern Cape Province, South Africa. *World Archaeology* 5:249-84.

_____. 1978. A preliminary report on the larger mammals from the Boomplaas Stone Age cave site, Cango Valley, Oudtshoorn District, South Africa. *South African Archaeological Bulletin* 33:66-75.

_____. 1984. Mammalian extinctions and Stone Age people in Africa. In *Quaternary Extinctions: A Prehistoric Revolution* (eds P. S. Martin and R. G. Klein): pp. 553-73. Tucson: University of Arizona Press.

Lewis-Willaims, J. D. 1984. Ideological continuities in prehistoric southern Africa: the evidence of rock art. In *Past and Present in Hunter Gatherer Studies* (ed C. Schrire): 225-252. Orlando: Academic Press.

Mazel A. and Parkington, J. 1978. Sandy Bay revisited: variability among Late Stone Age tools. *South African Journal of Science* 74:381-382.

Salinger, M. J. 1981. Palaeoclimates north and south. *Nature* 291:106-7.

Shackleton, N. J. 1973. Oxygen isotope analysis as a means of determining season of occupation of prehistoric midden sites. *Archaeometry* 15:133-41.

Spratt, D. A. 1982. The analysis of innovation processes. *Journal of Archaeological Science* 9:79-94.

Tusenius, M. 1986. *A study of charcoals from some archaeological contexts in South Africa*. (Unpublished M.A. dissertation, University of Stellenbosch).

Van Wyk R. 1979. Technological change: a macro perspective. *Technological Forecasting and Social Change* 15:281–96.

Vogel, J. C. 1983. Isotopic evidence for past climates and vegetation of South Africa. *Bothalia* 14:391–394.

Von den Driesch, A. and Deacon, H. J. 1985. Sheep remains from Boomplaas Cave, South Africa. *South African Archaeological Bulletin* 40:39–44.

Webley, L. 1978. *Sediment analysis of Boomplaas Cave*. (Unpublished B.A. Hons. thesis: University of Stellenbosch).

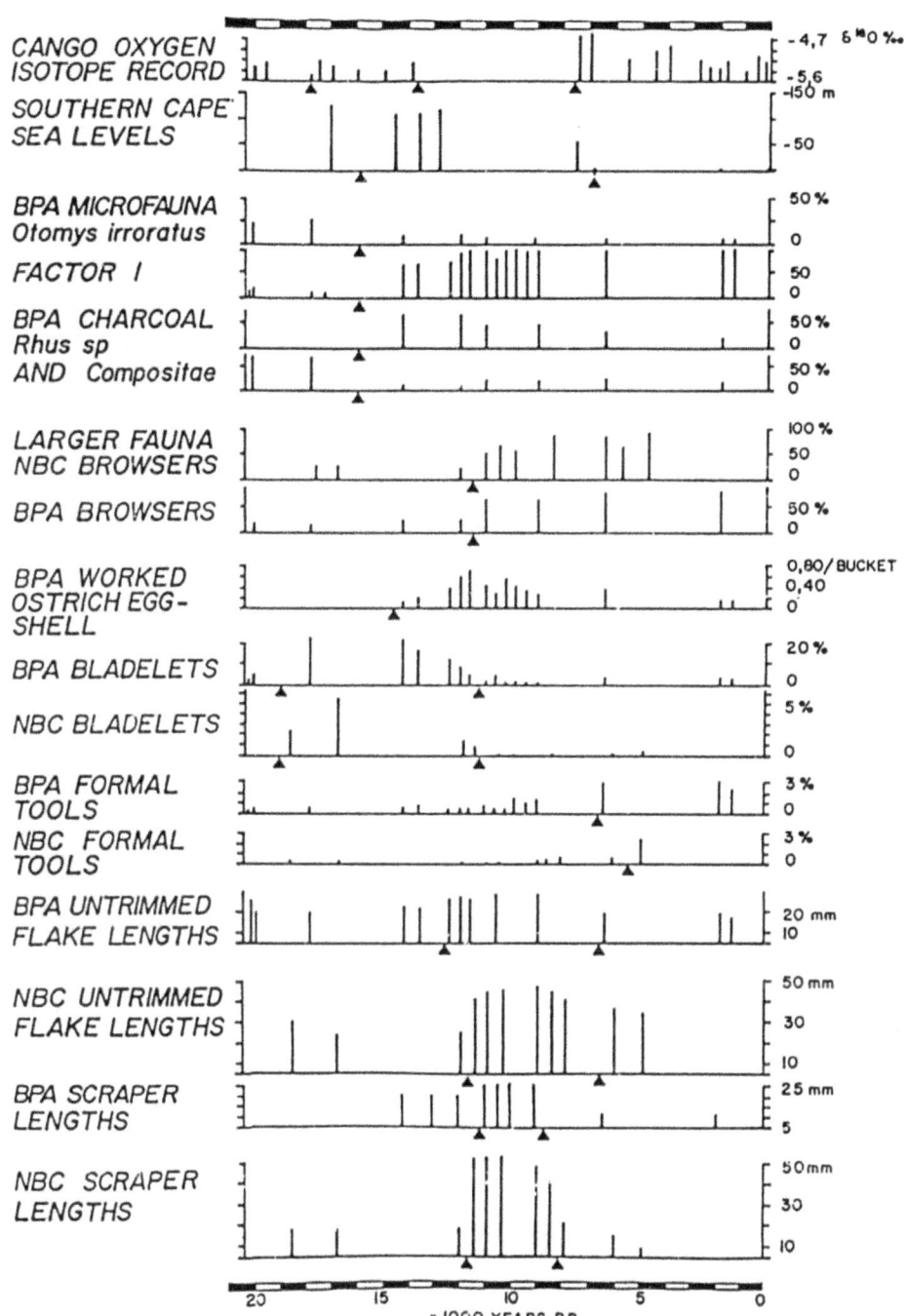

Fig. 9.1: A comparison of the timing of changes in various environmental and technological parameters at Boomplaas Cave (BPA) and Nelson Bay Cave (NBC). Cango oxygen isotope record after Vogel (1983); southern Cape sea levels after Dingle and Rogers (1972); BPA microfauna after Avery (1982) and Deacon et al. (1984); BPA charcoal after Deacon et al. (1983); NBC larger fauna after Klein (1972a, b); BPA larger fauna after Klein (1978); all other data from J. Deacon (1984a). Triangles mark times of most obvious change.

Scale and Timing of Technological and Environmental Changes

Human Adaptation in Southern Africa During the Last Glacial Maximum

By Peter Mitchell, Donald Baden-Powell Quaternary Research Centre, Oxford, England.

Introduction

Research carried out in 1971 by P. Carter at the site of Sehonghong in the Drakensberg mountains of Eastern Lesotho has shown that this area was occupied by Later Stone Age (LSA) hunter-gatherers during and after the Last Glacial Maximum (LGM). Radiocarbon dates from the site bracket an early microlithic assemblage similar to the Robberg Industry of the Southern Cape Province (J. Deacon 1978, 1984a, b) between 17,820 ± 270 bp and 13,000 ± 140 bp (Carter and Vogel 1974; Carter 1978). Today the Drakensberg experience a seasonally harsh climate with cold winters accompanied by extensive frosts as well as by snow (Staples and Hudson 1938; Killick 1978). Paleoecological data document a generally colder and more severe climate than at present over most, if not all, of South Africa during the LGM (J. Deacon; Lancaster and Scott 1984) and stable isotope analysis of equid teeth from another Eastern Lesotho site (Melikane) provides local confirmation of this (Vogel 1983). This chapter sets out to develop a model which can account for the human occupation of such an area in terms of the broader distribution of sites within South Africa as a whole. The origin of early microlithic assemblages will also be discussed and suggestions made as to why this should seem to coincide with the onset of the LGM.

Early Microlithic Assemblages—Nature and Distribution

Archaeological research in southern Africa during the last twenty years has greatly altered earlier conceptions of the antiquity of the LSA. As recently as fifteen years ago, Clark (1970) could imply a correspondence between the beginning of the Holocene and the origin of the LSA, but it is now clear that LSA industries extend far back into the Late Pleistocene. Radiocarbon dates have been reported ranging from 33,000 to 12,000 bp for early microlithic LSA assemblages (J. Deacon 1984a:309), although the validity of claims made for very early dates at the sites of Border Cave (Beaumont 1978), Heuningneskrans (Beaumont 1981), and Rose Cottage Cave (Vogel and Beaumont 1972) can be disputed.

The chief distinguishing feature of these assemblages, with the exception of those from Border Cave and Heuningneskrans where bladelets are very few, is "the systematic production of small bladelets from standardized single-platform bladelet cores" (J. Deacon 1984a:310) using quartz, silcrete, and crypto-crystalline silicas such as chalcedony and agate. Very few of these bladelets show evidence of retouch or of macroscopic utilization and the frequency of formally retouched tools (e.g., backed microliths, scrapers, adzes) is extremely low (<1%). Despite this, "the LSA character of these late Pleistocene microlithic assemblages is shown in associated lithic and nonlithic artefacts that are directly comparable with items in Holocene assemblages and in the ethnographic present" (J. Deacon 1984a:310). These include ostrich eggshell beads and water containers, bone beads and polished bone points, decorated fragments of ostrich eggshell, bored stones, grindstones, and tortoise shell containers. The lithic assemblage from Sehonghong is definitely part of the same complex (Mitchell in prep.).

During the LGM and the period immediately leading up to it, i.e., from 25,000 to 16,000 bp, human occupation of South Africa seems to have been limited both quantitatively and spatially. Sites dated to this period are listed in Table 10.1. The definite geographical localization of these sites, along the rim of the interior plateau in the South-Western and Southern Cape Province, the Drakensberg mountains of Lesotho and the Eastern Transvaal, Natal, and Swaziland is shown in Fig. 10.1. No sites occur in the interior of the Republic of South Africa, with the single exception of Rose Cottage Cave close to the Lesotho-Orange Free State border. This same geographical pattern remains noticeable even in the period immediately following the LGM (i.e., from 16,000 to 12,000 bp) as both Table 10.2 and Fig. 10.2 show. Dated sites remain restricted to a rim running along the southern and eastern edges of the interior plateau, although the number of dated occupations does increase considerably. The only indication of an increase in the area of settlement is given by dates from the two sites of Dikbosch I and Wonderwerk, both in the Northern Cape Province, but in neither case does LSA occupation extend back beyond 13,500 bp (Humphreys 1979; A. Thackeray 1981).

It would be incorrect to imply that the interior of South Africa was *not* inhabited by human populations during the LGM. Following Deacon and Thackeray (1984), we can, however, suggest that the spatiotemporal distribution of radiocarbon dates reflects the density of population over the subcontinent if it can be shown that the distribution in question is not a function of the localization of archaeological research.

Fig. 10.3, adapted from Parkington (1984), illustrates the location of research programmes carried out in LSA archaeology over the last twenty years. It is true that there is a relation between the relatively large number of sites in the Southern Cape and the extensive field work carried out there by H. and J. Deacon (H. Deacon 1976; H. Deacon and Brooker 1976; J. Deacon 1984a), Klein (1974), and Schweitzer (Schweitzer and Wilson 1981). But it is equally true that other comparatively well-researched areas, such as the Northern Cape Province (Humphreys

1979; A. Thackeray 1981) and the Middle Orange River Valley (Sampson 1972, 1985), have signally failed to provide evidence for human occupation during the Upper Pleniglacial (25,000 to 16,000 bp).

Table 10.1: Archaeological assemblages from South Africa dated to the Upper Pleniglacial 25,000–16,000 bp

Site	Date	Associated Assemblage	Reference
Heuningneskrans	24,630 ± 300	Early microlithic LSA	Beaumont 1981
	23,900 ± 800	Early microlithic LSA	Beaumont 1981
	21,100 ± 300	Early microlithic LSA	Beaumont 1981
	20,500 ± 300	Early microlithic LSA	Beaumont 1981
Sibebe Shelter	22,850 ± 160	MSA	Price-Williams 1981
Shongweni	22,990 ± 310	Early microlithic LSA	Davies 1975
Sehonghong	20,900 ± 270	MSA	Carter & Vogel 1974
	20,240 ± 230	MSA	Carter & Vogel 1974
	19,860 ± 220	MSA	Carter & Vogel 1974
	17,820 ± 270	Early microlithic LSA	Carter 1978
Melikane	20,000 ± 170	Postdating MSA occupation	Carter 1978
Nelson Bay Cave	18,660 ± 110	Robberg Industry	Deacon, J. 1984b
	18,100 ± 550	Robberg Industry	Deacon, J. 1984b
	16,700 ± 240	Robberg Industry	Deacon, J. 1984b
Boomplaas A	21,220 ± 195	Early microlithic LSA	Deacon, J. 1984b
	21,110 ± 420	Early microlithic LSA	Deacon, J. 1984b
	21,070 ± 180	Early microlithic LSA	Deacon, J. 1984b
	17,830 ± 180	Robberg Industry	Deacon, J. 1984b
Buffelskloof	22,800 ± 850	Unnamed Industry	Opperman 1978
	22,575 ± 270	Unnamed Industry	Opperman 1978
Elands Bay Cave	20,180 ± 220	Unnamed Industry	Parkington 1977

Note: At Rose Cottage Cave an early microlithic assemblage is present in levels postdating 25,640 ± 220 bp (Beaumont 1978).

Certainly caves, which might have acted as repeated foci of settlement and which could have preserved extensive sequences of occupation, are, on the whole, rare in such areas compared with the Cape Fold Mountain Belt and the Drakensberg. Most sites in these areas are consequently surface assemblages, generally consist only of lithic artefacts and are undateable by isotopic means. There is a problem of recognition as many early microlithic surface sites may only represent a limited range of activities and/or a transient occupation, and a lithic occurrence restricted to flakes, bladelets, and possibly cores is likely to be difficult to distinguish from later Holocene assemblages. Springs, pans, and similar features could, however, be expected to have repeatedly acted as foci of occupation and thus to have seen the buildup of a larger, more representative, and hence more easily characterizable assemblage. Nevertheless, while assemblages of both Smithfield and Wilton character are well known, and while Middle Stone Age (MSA) and Early Stone Age (ESA) occurrences are also common, *no* assemblages of early microlithic kind have been reported from such well-surveyed areas as the Seacow Valley (Sampson 1985). Given the ample documentation in the ethnographic record of a fairly direct relation between the density and distribution of hunter-gatherer populations on the one hand and the availability, predictability, and reliability of subsistence resources on the other (Birdsell 1953; Lee and De Vore 1968; Netting 1971), it seems reasonable to attempt to explain in ecological terms the late Pleistocene distribution of sites in southern Africa.

South Africa's Climate During the Upper Pleniglacial

A wide range of proxy data has been used to develop a picture of climatic and environmental conditions in South Africa during the late Pleistocene. Among the sources employed are studies of mammalian microfauna (Avery 1979), pollen (Scott 1982), charcoals (Scholtz 1985), sediments (Butzer 1984), oxygen-isotope analysis of speleothems (Talma *et al.* 1974) and high altitude periglacial features (Harper 1969). The relevant data have recently been summarized (J. Deacon, Lancaster, and Scott 1984). These studies are in agreement with the results of the CLIMAP project (CLIMAP 1976) and with global general circulation models (Gates 1976; Manabe and Hahn 1977; Heath 1979) indicating that, on the whole, conditions were cooler and drier worldwide during the LGM.

General circulation models agree in suggesting that during the LGM temperatures were depressed over South Africa by between 4° and 6° C. Local paleoclimatic data from the sources just mentioned confirm that conditions were distinctly colder, not only in the Cango Valley in the Southern Cape but also in the Transvaal and in Lesotho. Oxygen-isotope analysis of a speleothem from the Cango Caves has provided an estimate of a 5° C temperature depression at the LGM (Vogel 1983) and this is confirmed by a similar study of groundwater aquifers from near Uitenhage (Heaton 1981; Vogel 1983).

As with temperature, global general circulation models and terrestrial paleoclimatic data agree that LGM rainfall patterns differed considerably from those of the present. Nevertheless, 13-C isotope analysis of equid teeth from the site of Apollo 11 Cave in Southern Namibia suggests "that the winter rainfall area along the west coast was not extended appreciably further northwards during the Last Ice Age" (Vogel 1983:393). Earlier models (e.g., Van Zinderen Bakker 1976), which posited a straightforward northward displacement of the winter rainfall zone, must therefore be rejected. What remains at issue is the extent to which precipitation was reduced over all or only part of South Africa.

Table 10.2: Archaeological assemblages from South Africa dated to the Late Glacial 16,000–12,000 bp

Site	Date	Associated Assemblage	Reference
Heuningneskrans	13,100 ± 110	Early microlithic LSA	Beaumont 1981
	12,590 ± 130	Early microlithic LSA	Beaumont 1981
Siphiso Shelter	13,000	"Lebombo Industry"	Price-Williams & Prior 1985
Shongweni	14,760 + 130	Early microlithic LSA	Davies 1975
Sehonghong	13,000 ± 140	Early microlithic LSA	Carter & Vogel 1974
Melkhoutboom	15,400 ± 120	Robberg Industry	Deacon, H. 1976
Kangkara	12,550 ± 110	Robberg Industry	Deacon, J. 1984b
	12,330 ± 130	Robberg Industry	Deacon, J. 1984b
Boomplaas A	14,200 ± 240	Robberg Industry	Deacon, J. 1984b
	13,210 ± 55	Robberg Industry	Deacon, J. 1984b
	12,480 ± 130	Robberg Industry	Deacon, J. 1984b
	12,060 ± 105	Robberg Industry	Deacon, J. 1984b
Byneskranskop 1	12,730 ± 185	Robberg Industry	Schweitzer & Wilson 1982
Elands Bay Cave	12,450 ± 280	Robberg Industry	Parkington 1977
Wonderwerk	12,380 ± 95	Albany Industry	Humphreys & Thackeray, A. 1981
	12,160 ± 115	Albany Industry	op. cit.
	12,130 ± 110	Albany Industry	op. cit.
Dikbosch 1	13,770 ± 130	Unnamed industry	Humphreys 1979
	13,510 ± 120	Unnamed industry	Humphreys 1979
	13,240 ± 125	Unnamed industry	Humphreys 1979
	12,450 ± 100	Unnamed industry	Humphreys 1979

Note 1: Dates from Bushman Rock Shelter of 12,950 ± 40, 12,800 ± 75, 12,510 ± 105, 12,470 ±120, 12,310 ± 120, 12,160 ± 95 and 12,090 ± 95 bp are inconsistent with others from the same levels of 27,400 ± 1600 and 32,900 ±1800 bp and the assemblage may be mixed (Plug 1981). Re-analysis by L. Wadley (personal communication) does, however, suggest that an early microlithic LSA assemblage is present.

Note 2: An early microlithic assemblage very similar to that from Sehonghong comes from the site of Ravenscraig and is dated 10,200 ±100 bp, a date believed for typological and stratigraphic reasons to be too young (Opperman 1984). It may thus also be of Late Glacial date.

J. Deacon (1984a:35), in a discussion of the paleoenvironmental evidence for the Southern Cape, indicates that cooler ocean temperatures *ca.* 18,000 bp, coupled with weakening of the Agulhas Current (both of which are attested by studies of deep sea cores—e.g., Prell *et al.* 1979), would have reduced orographic precipitation in the area. The exposure of the continental shelf, which would have placed the Southern Cape mountains some 70 to 90 km. inland, and the reduction of moisture carried by westerly winds because of cooler ocean temperatures, would have increased this effect. A fall in mean annual temperature of 6° C could have decreased the amount of precipitable water by 30 to 50% along the coast, a figure which is in line with that inferred for the Southern Cape from Gates' (1976) general circulation model. Microfaunal, pollen, and charcoal data from Boomplaas A all support this by documenting markedly drier conditions during the LGM (H. Deacon *et al.* 1984). In the interior, pollen analyses of sediments from Wonderkrater in the Transvaal (Scott 1982) also indicate drier conditions described as "cooler temperate, dry subhumid or semiarid" during the LGM, although some problems exist with the site's dating (J. Deacon, Lancaster, and Scott 1984). Sedimentary evidence from Border Cave on the Natal/Swaziland border (Butzer *et al.* 1978a) indicates a cool to cold, possibly drier, environment from about 27,000 to about 13,000 bp, and a study of the mammalian microfauna by Avery (1982) confirms this interpretation.

Claims have, however, been advanced on geomorphological grounds for an increase in precipitation during the LGM in two areas of South Africa. In the South-Western Cape Butzer (1984a) has investigated dunes at Van Riebeeckstrand resulting from the marine regression of the LGM and has found them to be interrupted by three paleosol horizons which he attributes to three subhumid climatic phases between 25,000 and 15,000 bp. While there is no confirming biological evidence, this is consistent with the view expressed by Van Zinderen Bakker (1983:375) that "this region is dominated by the westerlies [and] during colder periods this influence will have been aggravated so that the southwestern Cape received more winter rain." Although precipitation is also argued to have increased in the Northern Cape Province during the LGM (Butzer *et al.* 1973; Butzer *et al.* 1978b) the evidence is limited and contradictory.

One further ecological factor of consequence to the distribution of human population is change in sea level during the LGM. A conservative estimate of an 85 m lowering at 18,000 bp has been offered by CLIMAP (1976), but data from the western and southern coasts of Africa (Dingle and Rogers 1972; Martin 1972) as well as from Barbados and New Guinea (Shackleton and Opdyke 1973) imply that a figure of closer to 120 or 130 m may be correct. The difference in coastal configuration that this will have produced is indicated in Fig. 10.4. Of particular interest is the finding that the ridge-and-vale nature of the Tertiary calcareous rocks underneath the area so exposed would have encouraged the formation of dunes and lagoons (Birch 1979) which are likely to have been the most important topographic features on the Agulhas Bank (J. Deacon 1984a:43). Such lagoons and marshes could have combined with the proximity of the coast to offer resources (shellfish, waterfowl) that might have been unobtainable further inland.

The LGM Paleoenvironment and the Distribution of Archaeological Sites

Fig. 10.5 illustrates the distribution of sites dating to the period 25,000 to 12,000 bp (i.e., the Upper Pleniglacial, including the LGM at 18,000 bp and the Late Glacial) in relation to the present temperature regime of South Africa. From this it can be seen that sites tend to concentrate in colder areas with mean annual temperatures ranging at present from 18° to 12° C. Temperature variation on a regional scale is largely controlled by two factors, namely altitude and latitude (Wellington 1955:219), neither of which can be expected to have been significantly different from the present. The same *relative* pattern of colder and warmer areas may therefore be inferred for the Pleistocene. It follows then that human population density would seem to have been greatest in the colder parts of South Africa. This seems counterintuitive as extensive cold can be expected, and is known to have had a deleterious effect on the density, distribution, and diversity of animal and plant populations (H. Deacon and F. Thackeray 1984). What it signifies, however, is that some other factor or factors were probably more important in regulating population density.

One such factor is likely to have been the availability of water. This is the principal limiting factor for plant growth, and hence for plant and animal biomass, over most of South Africa today (Tainton 1981), and may also be expected to have been so during the LGM, even allowing for the effects of generally lower temperatures, increased frost frequency and intensity, and greater wind intensity. Fig. 10.6 illustrates the distribution of archaeological sites dating to the Upper Pleniglacial and Late Glacial in relation to present day mean annual precipitation. A gross correlation can be discerned between site distribution and those areas which experience moderate to high rainfall. The problem is how this pattern can be transformed qualitatively to approximate that of the LGM.

It is apparent from earlier remarks that during the Upper Pleniglacial rainfall decreased over the Southern Cape, Natal and parts of the Transvaal, and indeed probably did so over most of South Africa's interior, and that the boundary between summer and winter rainfall zones did not shift significantly. Increased precipitation is possible, but by no means certain, over the South-Western and Northern Cape. Lowered temperatures will of course have diminished evaporation, but it seems improbable that this could have balanced the concomitant drop in precipitation (Sarnthein 1978:46), except perhaps in desert or semi-desert areas, such as the Southern Namib, where an increase in effective precipitation may be expected to have been associated with greater plant productivity.

In the present winter rainfall area precipitation is associated with the easterly movement of cyclones and anticyclones originating in the South Atlantic Ocean (Wellington 1955:259). Orographic effects and the proximity of the Cape Fold Mountain Belt to the sea result in "the high concentration of rainfall on the mountains backing the coast and the limitation of the heavier lowland rainfall to the coastal forelands," while "to the leeward of the mountain ranges the rain shadow is accompanied by semi-aridity and xerophyllous vegetation" (Wellington 1955:240). In the summer rainfall region, where precipitation derives mainly from the inflow of oceanic air streams from east coast highs, orographic effects (caused here by the Drakensberg mountains) and distance from the Indian Ocean also combine to render Natal, Swaziland, Eastern Lesotho, and the Eastern Transvaal the wettest parts of South Africa (Wellington 1955).

Allowing for most (or possibly all) of South Africa to have experienced a drier climate than that of today during the Upper Pleniglacial, orographic effects and distance from the sea are still likely to have resulted in areas such as Natal and those on the windward side of the Drakensberg and the Cape Fold Belt having been the relatively wettest parts of the country.

A Tentative Ecological Model of LGM Site Distribution

Rainfall and temperature, together with other factors not discussed here such as humidity, wind velocity, and evaporation rate, provide the climatic framework within which the resources exploited by a hunting-gathering population exist. Although the availability of shelter and of inorganic raw materials, particularly isotropically fracturing stone, are of some importance in determining the choice of particular settlement loci, it is the productivity of animal and plant resources, coupled with the availability of fresh water, that principally influence the density of population on a larger scale (Jochim 1976).

Direct measures of paleoproductivity are not obtainable, but estimates can nonetheless be made from correlations observed among contemporary data. Thus, Whittaker (1970:82) demonstrates that a nearly linear positive correlation exists between mean annual precipitation and net primary productivity in areas receiving up to 800 mm of rain a year, although differences in vegetation type may modify the relationship (F. Thackeray 1977). Similarly, Coe et al. (1976) have found that there is a statistically significant positive correlation between herbivore biomass on the one hand and mean annual precipitation on the other in areas receiving up to 700 mm of rain a year; it is also possible that this correlation may remain valid in areas which obtain up to 1400 mm of precipitation a year (cited in F. Thackeray 1977). As a first generalization it may therefore be suggested that the relatively wetter regions, in which sites dating to the LGM occur, were also the most ecologically productive parts of South Africa.

Studies of rainfall reliability and drought occurrence have shown that the wetter parts of South Africa tend to be those which suffer least from drought (Wellington 1955:255) and that rainfall predictability is in general strongly positively correlated with mean annual precipitation (Harris 1980). The overall effect of this will have been to increase animal and plant productivity and thus enhance further the security of hunter-gatherer populations in areas receiving higher precipitation compared with those receiving less.

Reexamination of Fig. 10.6 will show that, even within those areas likely to have been relatively wettest during the LGM, some regions are apparently devoid of dated archaeological sites and may therefore have supported a smaller human population. The eastern half of the Orange Free State and the Transvaal west of the Drakensberg mountains are examples of such areas. In explaining this, attention should be paid to a factor mentioned by J. Deacon (1984b:322), namely "that the distribution of sites of this time range in the Cape ecozone and the eastern half of southern Africa suggests that regions of moderate topographic variability...were preferred." The apparent concentration of sites in areas with greater variability in relief, even within the wetter parts of South Africa, suggests that topographic diversity itself offered some advantages for the maintenance of human populations under LGM conditions.

In discussing the density of Holocene populations Humphreys (1979) has drawn attention to the "unstructured" nature of environments in the interior of South Africa which would have resulted in this being a region marginal for human settlement under particularly severe conditions such as those of the LGM because it lacked "the diversity and concentration of reliable resources to maintain communities within defined home ranges through the seasonal round" (H. Deacon and F. Thackeray 1984:385). Lee (1979) has further pointed out that it is necessary to consider hunter-gatherer adaptations not just in the context of one year but over several decades. Access to the variety of resources required to support a population under rare but extremely adverse conditions is thus of greater importance than their short-term abundance in areas which are marginal over the longer term. It may be suggested that the unstructured environments of the interior plateau were insufficiently varied to offer that diversity of resources which could have provided a secure basis for long-term, large-scale human population maintenance.

Increasing topographic diversity on the other hand could be expected to have favoured a greater diversity of habitats, and hence of exploitable resources, and to have concentrated this diversity within a comparatively small area. Harpending and Davis (1977) show on theoretical grounds that the size of a group's subsistence territory will vary inversely with environmental diversity. A tentative indication of inter-regional differences in ecological diversity may be given if we consider the distribution of veld types in South Africa as described by Acocks (1975). Variety in the number of different veld types occurring within same-sized areas in different parts of the country is due to variation in climate, topography, and lithology, and as already stated climatic variations are governed by factors such as altitude and topography which are likely to have been important even under the very different conditions of the LGM. Areas showing great variation in the number of veld types present today may therefore also have been more diverse in the past. Examination of Table 10.3 will show that areas in the environs of Heuningneskrans in the Eastern Transvaal and Boomplaas A in the Southern Cape, for example, are more diverse than those in the Northern Cape or the Middle Orange River valley.

Table 10.3: Variation in ecological diversity within site territories (radius = 50 km) as measured by variation in the number of different veld types within them (after Acocks 1975).

Site	Number of different veld types within site territory
Melkhoutboom	8
Heuningneskrans	6
Boomplaas A	5
Sehonghong	4
Wonderwerk	2
Middle Orange River Valley	1

Large migratory and nomadic ungulates form an important part of the faunal samples from the sites under consideration (Klein 1978a, b). It is possible that these may have been easier to locate and hunt successfully in situations where broken topography resulted in fragmentation of grazing resources and impeded, or exercised a controlling influence over, their movements (Jacobson 1984) and where topographic features could have been of use in sighting and trapping prey.

Ethnographic data derived from studies of modern San indicate that "in energy returns hunting is a less rewarding activity than gathering, and vegetable foods provide the major part of the diet" (Lee 1979:205). Worldwide, tropical, subtropical, and temperate hunter-gatherers have been observed to depend more upon collected foods, mainly plants but also small ground game and shellfish, than on hunting (Lee 1968). The reason for this is that while plant collecting is a relatively high-yield and low-risk activity (since plants are immobile, more predictable in occurrence, and generally safer and easier to exploit), hunting is a relatively low-yield but high-risk activity. As Webster (1981:577) points out, generalized subsistence strategies in which plants play a major role are also more efficient, both trophically and in terms of the energy costs associated with resource procurement and group movement, than more specialized ones which emphasize the hunting of large game. In fact, the principal importance of meat for most hunter-gatherers may be less to provide calories than to act as a protein supplement and to supply certain essential nutrients, such as fats, that cannot be obtained by other means (Hayden 1981:397).

There is no evidence that either the bow and arrow or the atlatl were available to hunter-gatherers in South Africa before the end of the Pleistocene and hunting weapons are therefore likely to have been limited to an unaided throwing or thrusting spear. If anything, this should have resulted in an even lower success rate for hunting than that recorded ethnographically by Lee (1979) and Silberbauer (1981a) for modern San. As Hayden (1981:374) points out, social mechanisms, such as an increase in group size accompanied by pooling of kills and their sharing among all group members, a practice well documented for the San, could have gone some way to counteract this, as could the increased use of alternative hunting methods such as pit-traps. Nevertheless, the relative unreliability of hunting as a major subsistence strategy is not gainsaid. The significance of this is that an abundance of large game within an area, particularly of migratory or nomadic species, may by itself have been insufficient to have rendered an area suitable for maintaining a human population. The existence of more stationary and reliable resources, such as plant foods, ground game and shellfish, would then have been critical.

On a broad level plants can be divided into three categories (Grimes 1977):

1) those with underground storage organs, which tend to dominate in arid environments with less than 500 mm of rainfall a year;

2) grasses, which can more quickly respond to increases in the availability of water and nutrients and which tend to dominate in

areas with mean annual precipitations of between 500 and 1500 mm per annum;

3) trees and other herbaceous vegetation, which tend to dominate when an area receives a rainfall of more than 1500 mm a year.

Foley (1982), who first drew attention to the importance of this classification as an aid in the modelling of hunter-gatherer economies, stresses that the second category of grass-dominated communities, although they are associated with a high animal biomass, are "unsuitable for direct human consumption." The highveld of the interior plateau is believed to have been covered by grassland environments during the LGM (Van Zinderen Bakker 1982) and this may therefore be one further reason in explaining the sparse occupation of the region by hunter-gatherer groups.

Direct evidence for plant utilisation by Late Pleistocene hunter-gatherers in South Africa is extremely limited, but the occurrence of edible seeds and leaves at Border Cave (Beaumont 1978), and possibly also at Heuningneskrans (Beaumont 1981) as well as of so far unanalysed carbonised remains at Boomplaas A and Elands Bay Cave (J. Deacon 1984a:312) hints at their use. While it is important not to place unequivocal reliance on the contemporary ethnographic record and thus refuse to admit the likelihood of substantial differences between the lifeways of the modern San and those of their prehistoric ancestors, it is equally important not to be blinded by the survival of faunal remains in archaeological assemblages into thinking that Late Pleistocene populations depended solely upon "big-game" hunting. One extremely important resource known to have been exploited by later Holocene groups is geophytic plants such as *Cyperus*, *Hypoxis*, *Moraea*, and especially *Watsonia* (H. Deacon 1976). These plants tend to occur on nutrient-poor soils, for example, within the fynbos biome of the Southern Cape, and are extremely productive, though taking a long time to reach maturity (J. Deacon 1984b:258-59). Furthermore, their productivity increases substantially after firing of the landscape. This seems to have been a common practice among Holocene hunter-gatherers and the fact that the controlled use of fire (as evidenced by the presence of hearths in MSA contexts) dates back some 100,000 years or more in the Southern Cape must leave open the possibility of a great antiquity for such "firestick farming" (H. Deacon 1985). Such geophytic plants may well have been an important resource to Late Pleistocene groups, although their abundance is likely to have been reduced as the fynbos was fragmented by grassland expansion.

It can also be suggested that the resources of the coast and coastal lagoons may help account for relatively dense human settlement in the

Late Pleistocene of southern Africa. Human exploitation of shellfish is already attested in MSA deposits at Klasies River Mouth in layers probably dating back beyond 125,000 bp (Singer and Wymer 1982), and there is no reason why this resource should not also have been used during the Last Glacial, even though evidence of this will have been lost by the post-glacial rise in sea level. As indicated by their use in other parts of the world (Bailey 1975), shellfish may have been a supplemental resource at any season or a primary one when other, high-return resources were temporarily unavailable. Their possible significance in supporting coastally oriented populations ought not to be ignored, as Meehan, referring to an Australian example, points out:

> Because shellfish were collected consistently, they provided a small, constant source of fresh protein, the importance of which should not be underestimated. Certainly one wallaby provided a larger quantity of flesh, protein and energy than a single haul of shellfish, but even in good seasons such large mammals may only have been available every four or five days at most (1977:256).

The model which has been developed here suggests that the distribution of human populations in South Africa during the Upper Pleniglacial, including the LGM, can be understood in terms of the effects which precipitation and topographic diversity are likely to have had on the availability, reliability, and productivity of subsistence resources. Sites are concentrated in areas of South Africa which were probably relatively wetter even under LGM conditions, and within these an emphasis on areas marked by considerable variability in relief and ecology can be discerned. It is in these areas that resource productivity and diversity are likely to have been highest. The model therefore substantiates the pattern seen in the distribution of dated archaeological sites (Fig. 10.1 and Table 10.4), and in particular strengthens the case against intensive occupation of the interior at this time. Thus, although the converse has recently been argued, namely that "the Robberg assemblages and associated faunal remains are occasional, peripheral and perhaps atypical reflections of a system centred elsewhere, in the interior for example," (Parkington 1984:120), ecological considerations imply strongly that the Southern Cape was one of the few regions of South Africa productive enough to have supported a substantial hunter-gatherer population under Last Glacial conditions. Previously, it has been suggested that human adaptive strategies during the LGM saw relatively more time invested in obtaining large, migratory ungulates and relatively less in exploiting underground plant foods and territorial and ground game than was the case in the later Holocene (H. Deacon 1972) However, it is suggested here that predictable, immobile resources, principally plant foods, will still have played a fundamental role in subsistence strategies and that this will have favoured the concentration of settlement into the more productive and ecologically diverse areas along the rim of the interior plateau and along the coast. A broadly based adaptive strategy, marked by considerable behavioural flexibility and only loose forms of territoriality (Yellen 1977) will have offered the best means of coping with the reduced quantity and diversity of foods characteristic of harsh environments.

Table 10.4: List of site names referred to in Fig. 10.1 and 10.2.

Abbreviation	Site name	Reference
BNK	Byneskranskop 1	Schweitzer & Wilson 1982
BPA	Boomplaas A	Deacon, H. & Brooker 1976; Deacon, J. 1984b
BRS	Bushman Rock Shelter	Plug 1981
BUF	Buffelskloof	Opperman 1978
DIK	Dikbosch 1	Humphreys 1979
EBC	Elands Bay Cave	Parkington 1977; 1980
HNK	Heuningneskrans	Beaumont 1981
KRA	Kangkara	Deacon, J. 1984b
MHB	Melkhoutboom	Deacon, H. 1976
MLK	Melikane	Carter 1978
NBC	Nelson Bay Cave	Deacon, J. 1978; 1984b
RAV	Ravenscraig	Opperman 1984
RC	Rose Cottage Cave	Beaumont 1978
SH	Sehonghong	Carter 1978
SHO	Shongweni	Davies 1975
SIB	Sibebe Shelter	Price-Williams 1981
SIP	Siphiso Shelter	Price-Williams & Prior 1985
WK	Wonderwerk	Thackeray, A. 1981

An Ecological Model for the Adoption of a Microlithic Technology

While use of a microlithic technology has long been employed as a criterion in the definition of the "Later Stone Age" (Goodwin and Van Riet Lowe 1929), relatively little attention has been paid to the origins of this technology and the advantages which it may have offered over that of the MSA. The traditional emphasis on LSA lithic typologies (e.g., Sampson 1974) is likely to have been partly responsible for this, as is the prevalence of models interpreting changes in lithic traditions in terms of diffusion, migration, and differences in the availability of appropriate raw materials (see Parkington 1984 for a discussion of this). In this respect, Clark's (1959:169–72) consideration of the advantages of using finer-grained raw materials and of manufacturing barbed artefacts, and Phillipson's (1976, 1979) discussion of the uses of backed microliths stand out as functional arguments. So far, however, attention has not been given to the ecological framework within which LSA technology first appeared.

Little convincing evidence exists for microlithic assemblages earlier than 25,000 bp. At Heuningneskrans, the earliest well-established radiocarbon date on microlithic material is 24,630 ± 300 bp. Earlier dates rest on a backward extrapolation of the rate of deposit accumulation and on potentially unreliable amino-acid racemization dates, while the character of the lithic assemblage cannot be assessed in the absence of its detailed study and publication (Beaumont 1981). The so-called "Pre-Wilton" assemblage from Rose Cottage Cave is associated at its base with two inverted radiocarbon dates of 25,640 ± 220 and 29,430 ± 520 bp (Vogel and Beaumont 1972; Butzer 1984a) which, in the absence of an accurate stratigraphy for the site, can only be suggested to provide *termini post quos* for this occurrence. At Border Cave the "Early LSA" assemblage shares few of the features said to characterize this complex (Beaumont 1978) and associated radiocarbon dates of 45,000 to 35,000 bp may be misleading as micromammalian faunal samples would suggest contemporaneity with the LGM as represented at Boomplaas A (Avery 1982; H. Deacon, personal communication). Finally, L. Wadley (personal communication), in the absence of further determinations, is unwilling to accept unreservedly a date of greater than 29,000 bp associated with an Early LSA assemblage at Cave James in the Magaliesberg mountains of the Western Transvaal.

More securely dated assemblages suggest that the latest MSA industries in South Africa date to rather later than 32,400 ± 420 bp and somewhat earlier than 21,220 ± 160 bp at Boomplaas A (H. Deacon and Brooker 1976; J. Deacon 1984a), to 22,850 ± 160 bp at Sibebe Shelter in Swaziland (Price-Williams 1981), to 26,300 ± 400 bp at Apollo 11 Cave in Southern Namibia (Wendt 1976), and to before *ca.* 20,000 bp at Melikane and Sehonghong in Eastern Lesotho (Carter and Vogel 1974; Carter 1978). Thus the end of the MSA industries and the beginning of LSA ones appear to fall between 30,000 and 20,000 bp, and may well more nearly coincide with the climatic deterioration to the cold, dry, and generally harsh conditions of the LGM which began about 25,000 bp (H. Deacon *et al.* 1984). This is particularly well demonstrated at Boomplaas A where the earliest LSA assemblages from members LP, LPC, and YOL coincide with a sharp decline in precipitation and temperature (Avery 1979).

In a consideration of the relation between technology and environment Torrence (1983) has drawn attention to what she has termed "time-stressed" environments, i.e., ecologically specialized zones which have a limited range of relatively mobile and unpredictable resources. Diversity indices for a number of environmental indicators, such as charcoals, pollen, and microfauna, demonstrate beyond doubt that the LGM was a period of markedly reduced ecological diversity and productivity compared with both the Inter-Pleniglacial and the Holocene (H. Deacon *et al.* 1984). In such an environment acquisition of subsistence resources may be at a premium. The efficient use of time will be favoured by such behaviours as the scheduling of resource exploitation and by increasing the speed and efficiency of critical activities such as pursuing and capturing prey (Torrence 1983). Two further suggestions can be added to this model. Reducing the time invested in raw material procurement by expanding the range of materials that could be exploited would also have increased the time that could be devoted to resource procurement. Decreasing the time required for artefact maintenance and manufacture would have led to the same result, and it may be noted that ethnographic evidence (e.g., Silberbauer 1981b:482) suggests that such activities can be particularly time-consuming among modern hunter-gatherers.

How does the adoption of a microlithic technology relate to such considerations? Clarke (1976:461) has identified four principal advantages which such a technology possesses over a nonmicrolithic one. In the first place, it allows the maximum length of cutting edge and the maximum number of utilisable pieces of stone to be extracted from a given volume of raw material. Bordaz (1970) suggests that this represents a 1000% increase in efficiency over an Upper Palaeolithic (cf. MSA) blade-based technology.

Secondly, a microlithic technology permits regular exploitation of small, nodular pebbles for the manufacture of even large artefacts via the development of more complicated composite tools. Related to this is the further advantage of allowing more effective exploitation of small sources of extremely hard and/or sharp stone such as agate and chalcedony. One consequence of these factors may have been to facilitate the exploitation of territories lacking, or poor in, alternative raw materials.

Finally, while the MSA toolkit almost certainly included composite artefacts such as wooden-hafted, stone-tipped spears (Inskeep 1978:58), it became possible with a microlithic technology to construct "composite tools in terms of small, rapidly replaceable and interchangeable, standardized, and mass-produced units, manufactured in advance and kept in readiness for inevitable wear-and-tear—a pull-out and plug-in construction" (Clarke 1976:461). Referring to European examples, Clarke further points out that while "a broken Solutrean spearhead or a splintered Magdalenian harpoon required a complete, specially designed replacement [in] most composite tools the breakage would be localized to one or two elements" alone (Clarke 1976). Concentrations of bladelets found in Robberg Industry levels at Boomplaas A suggesting that they may have been kept in some kind of container (J. Deacon 1982), possibly for transport from place to place, would be consistent with this aspect of Clarke's hypothesis, as would the very large number of bladelets

P. Mitchell

characteristic of Robberg assemblages—these could have been made in preparation for future use and fitted either at residential sites or at other, task-specific loci, such as hunting stations. Restriction of their manufacture to only some sites has the important implication of rendering it difficult to identify the other components of the Robberg settlement system since, as already mentioned, early microlithic assemblages lacking bladelets and bladelet cores may be indistinguishable from other LSA occurrences.

As already implied, the bladelets characteristic of early microlithic LSA assemblages must have been hafted. Their small size (length under 25 mm, and width of less than 10 mm) leaves no doubt about this, although direct evidence such as traces of mastic or polish (Keely 1982) is lacking. The absence of bone hafts from the assemblages excavated so far is unlikely to be due to problems of preservation since both faunal remains and other bone artefacts are found. It should therefore be ascribed to an exclusive use of wooden hafts and/or to extensive curation with loss tending to occur away from the presumably residential cave sites (cf. Binford 1976). The advantages offered by hafting may have included the formation of cutting edges of a size and shape otherwise unobtainable and an increase in the force that could be exerted during work (Keeley 1982). An example of this would be the hafting of several bladelets in sequence to form the cutting edge of a knife. Microwear studies of bladelets from Holocene levels at Boomplaas A show that some at least were used in this way to cut wood (Binneman 1982).

It has been stated previously that the bow and arrow was probably unknown in South Africa during the Late Pleistocene and that an unaided stabbing or throwing spear may therefore have been the most advanced projectile weapon available. The use of a barbed spear would go some way toward accounting for the large number of bladelets present in early microlithic assemblages. As ethnographic evidence demonstrates (Flood 1983:189–90), such a weapon could have increased both the depth of a wound and the loss of blood from it, while the barbs could have helped to hold the spear in the animal and thus impede its escape (Hayden 1979:82). If poison were also used, then holding the spear in the animal will have facilitated its entry and spread through the circulatory system (Clark 1959:171–2). The effectiveness of spear-hunting should not be disparaged by the contemporary reliance of the San on the bow and poisoned arrow. Ethnographically the use of poisoned stabbing and thrusting spears is well attested among them, whether to kill large game at close quarters (Clark 1959:226) or to kill large animals such as eland, hartebeest, and kudu which have been run down by a group of hunters (Fourie 1928:99; Silberbauer 1981a).

The hypothesis put forward therefore has two components. First, a microlithic technology is advantageous in exploiting small, nodular materials more effectively, in maximizing the efficiency with which a given volume of stone can be exploited and in encouraging the manufacture of composite artefacts which can increase the force available during work, give greater flexibility to the length and shape of the cutting edge and are much easier to repair. The net effect of such advantages will have been to reduce the time required for raw material procurement and for artefact manufacture and maintenance and thus to increase the time that could be devoted to subsistence activities. This coupled with the increased effectiveness of barbed weapons for killing at close quarters are argued to have been selectively useful in the context of a time-stressed environment such as that of the LGM. The Boomplaas A sequence, which is the best studied and most detailed available, shows that an LSA technology was in use prior to the occurrence of assemblages which have been grouped together as the Robberg Industry. It is tempting to suggest, in the virtual absence of both utilised and formally retouched artefacts from this pre-Robberg assemblage, that hafted microlithic composite tools were already being produced and that the gradual increase in bladelet frequency, and, in the Robberg itself, the appearance of special core types (flat bladelet cores—J. Deacon 1984a:111), can be related to a shift from the use as microlithic inserts of unsystematically produced pieces of quartz to standardized, mass-produced bladelets. In terms of the hypothesis advanced here this would represent the development of a further economizing behaviour, still within the context of a time-stressed LGM environment.

Data from two other regions of the world may be useful to compare with this model. First, both the Dyuktai Culture of North-Eastern Siberia and the American Paleo-Arctic Tradition are characterized *inter alia* by wedge-shaped microblade cores and accompanying microblades (Morlan and Cinq-Mars 1982:373), the latter probably produced for the purpose of arming slotted antler or bone points (Anderson 1968). A recent review of earlier findings by Yi and Clark (1985:18) suggests that these industries may date back to 18,000–15,000 bp and could have developed within a tundra-steppe biome. Paleoecological data can be taken to indicate that resources may also have been difficult to exploit within such an area and that the environment could correctly have been described as "time-stressed." Without invoking environmental determinism, it can nevertheless be argued that in both Beringia and South Africa similar selective pressures in the direction of time-economizing behaviours and improved hunting technology may have helped to produce similar lithic technologies.

Also of interest is Marks's (1983) discussion of the Middle-Upper Paleolithic transition in the Levant, where increasing aridity is thought likely to have necessitated increased group mobility. Such "increased mobility, with less and less security as to the predictability of available flint sources near any one site probably was met with additional 'economizing behaviour'" (Marks 1983:92). The net result of such changes as increased production of elongated blanks and a shift towards a specialized opposed platform reduction strategy was to reduce cores using only a single platform but with careful selection of raw material so that they were "preformed" to produce the elongated blanks (blades) typical of the Upper Paleolithic. In this case it is the time required for raw material procurement alone which is being minimized, but the principle is the same and suggests that other technological transitions might repay study along the lines developed by Torrence.

This model carries three important test implications for the archaeological record. The first of these for the origins of microlithic industries is that a trend towards increasing microlithization may be discernible for some time before a definitely LSA tradition is apparent. This is because South African environments can be expected to have become increasingly time-stressed over a period of several millenia prior to the LGM as a result of globally deteriorating climatic conditions. The progressive diminution of blade size recorded by Carter (1978) at Sehonghong and Moshebi's Shelter in Eastern Lesotho over a period dated to between 31,000 and 20,000 bp might provide preliminary support for such an idea.

Secondly, the transition to a microlithic technology may have first occurred in those parts of South Africa where environmental conditions were most time-stressed and economizing behaviours the most advantageous. It is tempting to suggest that such areas may have been those which provide least evidence for human occupation during the LGM. In other words, a microlithic technology may have first developed on the interior plateau and only later have spread to the surrounding Fold Mountain Belt/ Drakensberg region.

Lastly, where environmental conditions are becoming increasingly harsh over a wide area, innovations which represent economizing behaviours that can increase the time available for, and the efficiency of, food procurement may be readily diffused and adopted. The limited time frame for the transition from MSA to LSA technologies in southern Africa is congruent with this test implication.

Of previous attempts to explain the adoption of a microlithic technology in southern Africa, only Phillipson has developed a model which views this process within an ecological framework. He has argued that the advent of a LSA stone-tool tradition is correlated with a shift from large, gregarious prey towards the hunting of smaller game

(Phillipson 1976, 1979, 1984). The rationale for this is that the bow and arrow were adopted as more effective hunting weapons in the more densely vegetated environment which developed in the warmer and wetter conditions of the Late Glacial. Whatever the merits of this hypothesis in the Zambian case, it is clearly insufficient when applied to the South African data. Here the faunal record demonstrates quite clearly that early microlithic assemblages such as those at Boomplaas A and Nelson Bay Cave are associated with faunas dominated by equids and by alcelaphines and other large bovids characteristic of open country (Klein 1978a, b). Furthermore, Phillipson's hypothesis deals only with the introduction of backed microliths, not with that of a microlithic technology *per se*. In South African early microlithic assemblages backed and retouched microliths are almost completely absent (J. Deacon 1984a).

Conclusion

The aims of this chapter have been twofold. In the first place, it has tried to suggest an ecological framework within which the distribution of archaeological sites of the LGM in South Africa can be understood. Because of the lower productivity inferred for the much colder and drier environments of the LGM, high human population densities seem to have been restricted to the more favourable parts of the subcontinent. These seem to have been the relatively wetter areas with considerable topographic and hence ecological diversity where a broad range of animal and plant foods is likely to have been concentrated within the size of territory exploitable by pedestrian hunter-gatherers. The availability of plant resources, such as geophytes, may have been particularly important because of the probability that success rates for hunting large game were low. Coastlands may also have been a focus of relatively dense settlement with shellfish offering a relatively dependable and easily exploitable resource.

Examination of the distribution of sites dating to later periods suggests that the refuge areas of the LGM remained foci of human settlement for some time into the Late Glacial (16,000 to 12,000 bp). Palaeoecological evidence (J. Deacon, Lancaster, and Scott 1984) suggests a rapid rise in rainfall, but with cold conditions persisting during this period. Some increase in ecological productivity during the Late Glacial is expected and can be inferred from increasing charcoal, pollen, and micromammalian diversity indices at Boomplaas A (H. Deacon *et al.* 1984). The increase in the number of sites dated to this period compared with the Upper Pleniglacial further suggests a possible increase in the size of the human population. The continued concentration of sites along the rim of the interior plateau may be explained by postulating a lag effect in the expansion of population into previously more marginal areas. Immigration of groups from the continental shelf as world sea levels rose with the melting of the continental ice sheets must also be considered.

It is interesting to note that the mid-Holocene seems to have been another period of acute stress for South African hunter-gatherers, with the interior of both the Transvaal and the Orange Free State largely devoid of occupation between 9,000 and 4,500 bp (J. Deacon 1974). Areas such as the Southern Cape appear once more to have acted as refuge foci of dense human settlement (J. Deacon 1984b:322). Further back in time the periodicity of cold and warm, wet and dry phases throughout the Pleistocene suggests that regional pulses of occupation and non-occupation can be expected for ESA and MSA populations, too (H. Deacon and F. Thackeray 1984; Beaumont in press). A still unresearched question is the extent to which such discontinuities in population distribution may have influenced the physico-genetic differentiation of Khoisan populations from a broader, ancestral African base.

An attempt has also been made to consider the development of the LSA within the context of the selective advantages a microlithic technology may have offered in the time-stressed LGM environment, especially as regards raw material procurement, artefact manufacture, and maintenance and hunting efficiency. It remains to be noted that the early LSA includes not only a microlithic technology but also features a few artefacts in ostrich eggshell and in bone as well as evidence of art and of personal ornamentation (J. Deacon 1984a:310). This attests to the existence of broader, ongoing social processes which must also be addressed in building models of cultural change. The extent to which such processes were related to ecological developments is an issue which, though not addressed here, may have important implications for the origins of microlithic technology itself.

Acknowledgements

I am grateful for comments made to an earlier draft of this work by H. J. Deacon, P. L. Carter and M. Tusenius as well as by my supervisor, R. R. Inskeep. Responsibility for the final version naturally rests with me alone.

References

Acocks, J. P. H. 1975. Veld types of South Africa. *S. Afr. Bot. Surv.* Mem. 35

Anderson, D. D. 1968. A stone age campsite at the gateway to America. *Scientific American* 218 (6):24–33.

Avery, D. M. 1979 *Upper Pleistocene and Holocene palaeoenvironments in the Southern Cape: the micromammalian evidence from archaeological sites*. (Unpublished D. Phil. thesis, University of Stellenbosch).

Bailey, G. N. 1975. *The role of shell-middens in prehistoric economies.* (Unpublished Ph. D. thesis, University of Cambridge).

Beaumont, P. 1978. *Border Cave*. (Unpublished M. A. thesis, University of Cape Town).

———. 1981. Heuningneskrans. In *A Guide to Archaeological Sites in the Northern and Eastern Transvaal* (ed E. Voight): pp. 132–45. Pretoria: Southern African Association of Archaeologists.

———. In press. Where did all the young men go in 0-18 stage 2? *Palaeoecology of Africa.*

Binford, L. R. 1976. Forty seven trips: a case study in the character of some formation processes of the archaeological record. In *Contributions to Anthropology: the Interior Peoples of Northern Alaska*, (ed E. Hall): pp. 299–351. Ottawa: National Museum of Man Mercury Series 49.

Binneman, J. N. F. 1982. *Mikrogebruikstekene op steenwerktuie: eksperimentele waarnemings en 'n studie van werktuie afkomstig van Boomplaasgrot.* (Unpublished M. A. thesis, University of Stellenbosch).

Birch, G. 1979. Nearshore Quaternary sedimentation off the south coast of South Africa. *Marine Geoscience Research Group Report, University of Cape Town* 11:127–46.

Birdsell, J. B. 1953. Some environmental and cultural factors influencing the structuring of Australian Aboriginal populations. *American Naturalist* 87:17–200.

Bordaz, J. 1970. *Tools of the Old and New Stone Age*. Newton Abbot: David and Charles.

Bousman, B. 1985. On and off-site palaeoenvironmental research in the Blydefontein Basin. Unpublished paper delivered at the Southern

African Association of Archaeologists' Conference, Grahamstown, September.

Butzer, K. W. 1984a. Late Quaternary environments in South Africa. In *Proceedings of the International Symposium on Late Cenozoic Palaeoclimates of the Southern Hemisphere, Swaziland 1983* (ed J. Vogel): pp. 235-64. Rotterdam: Balkema.

Butzer, K. W. 1984b Archaeogeology and Quaternary environment in the interior of southern Africa. In *Southern African Prehistory and Palaeoenvironments* (ed R. G. Klein): pp. 1-64. Rotterdam: Balkema.

Butzer, K. W. and Helgren, D. M. 1972. Late Cenozoic evolution of the Cape coast between Knysna and Cape St. Francis, South Africa. *Quaternary Research* 2:143-69.

Butzer, K. W., Stuckenrath, R., Fock, G. J. and Zilch, A. 1973. Paleohydrology of a late Pleistocene lake, Alexandersfontein, Kimberley, South Africa. *Nature* 243:328-30.

Butzer, K. W., Beaumont, P. and Vogel, J. C. 1978a. Lithostratigraphy of Border Cave, Kwazulu, South Africa: a Middle Stone Age sequence beginning c. 195,000 bp. *J. Archaeol. Sci.* 5:317-41

Butzer, K. W., Stuckenrath, R., Bruzewicz, A. J. and Helgren, D. M. 1978b Late Cenozoic paleoclimates of the Gaap escarpment, Kalahari margin, South Africa. *Quaternary Research* 10:310-39

Carter, P. L. 1978. *The prehistory of Eastern Lesotho*. (Unpublished Ph. D. thesis, University of Cambridge).

_____. and Vogel, J. C. 1974. The dating of industrial assemblages from stratified sites in eastern Lesotho. *Man* 9:557-70

Clark, J. D. 1959. *The Prehistory of Southern Africa*. Harmondsworth: Penguin.

_____. 1970. *The Prehistory of Africa*. London: Thames and Hudson.

Clarke, D. L. 1976. Mesolithic Europe: the economic basis. In *Problems in Economic and Social Archaeology* (eds G. Sieveking et al): pp. 449-81. Cambridge: C. U. P.

CLIMAP. 1976. The surface of the ice-age earth. *Science* 191:1131-37.

Coe, M. D. *et al.* 1976. Biomass and primary productivity of African large herbivores in relation to rainfall and primary production. *Oecologia* 22:341-54.

Davies, O. 1975. Excavations at Shongweni South Cave. *Ann. Natal Mus.* 22:627–62.

Deacon, H. J. 1976. *Where Hunters Gathered: A Study of Holocene Stone Age People in the Eastern Cape.* Claremont, South African Archaeological Society. Monograph 1.

_____. 1985. Unpublished paper delivered at Fynbos Biome Project meeting, Stellenbosch, July.

_____. and Brooker, M. 1976. The Holocene and Upper Pleistocene sequence in the southern Cape. *Ann. S. Afr. Mus.* 71:203–14.

_____. and Thackeray, J. F. 1984. Late Pleistocene environmental changes and implications for the archaeological record in southern Africa. In *Proceedings of the International Symposium on Late Cenozoic Palaeoclimates of the Southern Hemisphere*, Swaziland, 1983 (ed J. Vogel): pp. 375–90. Rotterdam: Balkema.

_____., Deacon, J., Scholtz, A., Thackeray, J. F., Brink, J. S. and Vogel, J. C. 1984. Correlation of palaeoenvironmental data from the Late Pleistocene and Holocene deposits at Boomplaas Cave, southern Cape. In *Proceedings of the International Symposium on Late Cenozoic Palaeoclimates of the Southern Hemisphere*, Swaziland, 1983 (ed J. Vogel): pp. 339-51. Rotterdam: Balkema.

Deacon, J. 1974. Patterning in the radiocarbon dates for the Wilton/Smithfield complex in southern Africa. *S. A. A. B.* 29:3–18.

_____. 1978. Changing patterns in the late Pleistocene/early Holocene prehistory of southern Africa seen from the Nelson Bay Cave stone artefact sequence. *Quaternary Research* 10:84–111.

_____. 1982. *The Later Stone Age in the southern Cape, South Africa.* (Ph.D. thesis, University of Cape Town).

_____. 1984a. The Later Stone Age. In *Southern African Prehistory and Palaeoenvironments* (ed R. G. Klein): pp. 221-328. Rotterdam: Balkema.

_____. 1984b. *The Later Stone Age of Southernmost Africa.* Cambridge Monographs in African Archaeology 12. (BAR International Series 213).

_____, Lancaster, N. and Scott, L. 1984. Evidence for Later Quaternary climatic change in southern Africa. In *Proceedings of the International Symposium on Late Cenozoic Palaeoclimates of the Southern Hemisphere,* Swaziland 1983 (ed J. Vogel): pp. 391–404. Rotterdam: Balkema.

Dingle, R. V. and Rogers, J. 1972. Pleistocene palaeogeography of the Agulhas Bank. *Trans. Roy. Soc. S. Afr.* 40:155–65.

Flood, J. 1983. *Archaeology of the Dreamtime.* Sydney: Collins.

Foley, R. 1979. Incorporating sampling into initial research design: Soome aspects of spatial archaeology. In *Sampling in Contemprary British Archaeology* (eds C. Gamble, J. Cherry and S. Shennan). BAR International Series 50:49–66.

Fourie, J. 1928. The Bushmen. In *Native Tribes of South-West Africa* (ed C. L. Hahn): pp. 79–106. Cape Town.

Gates, W. L. 1976. The numerical sumulation of ice-age climate with a global general circulation model. *J. Atmosph. Sci.* 33:1844–73.

Goodwin, A. J. H. and Van Riet Lowe, C. 1929. The Stone Age cultures of South Africa. *Ann. S. Afr. Mus.* 27:1–289.

Grimes, J. P. 1977. Evidence for the existence of three primary strategies in plants and its relevance to ecological and evolutionary theory. *American Naturalist* 111:1169–94.

Harpending, H. and Davis, H. 1977. Some implications for hunter-gather ecology derived from the spatial structure of resources. *World Archaeol.* 8:275–86.

Harper, G. 1969. Periglacial evidence in southern Africa during the Pleistocene epoch. *Palaeoecology of Africa* 4:71–101.

Hayden, B. 1979. *Palaeolithic Reflections: Lithic Technology and Ethnographic Excavation among Australian Aborigines.* Canberra: Australia Institute for Aboriginal Studies.

_____. 1981. Subsistence and ecological adaptations of modern hunter-gatherers. In *Omnivorous Primates—Hunting and Gathering in Human Evolution* (eds R. S. Harding and G. Teleki): pp. 344–421. New York: Columbia Unversity Press.

Heath, G. R. 1979. Simulations of a global palaeoclimate by three different atmospheric general circulation models. *Palaeogeogr., Palaeoclimatol., Palaeoecol.* 26:291–303.

Heaton, T. H. E. 1981. Dissolved gases: Some applications to groundwater research. *Trans. Geol. Soc. S. Afr.* 84:81–97.

Humphreys, A. J. B. 1979. *The Holocene sequence in the northern Cape and its position in the prehistory of South Africa.* (Unpublished Ph. D. thesis, University of Cape Town).

_____ and Thackeray, A. I. 1983. *Gaap and Gariep*. (South African Archaeological Society Goodwin Series 4).

Inskeep, R. R. 1978. *The Peopling of Southern Africa*. Cape Town: David Philip.

Jacobson, L. 1984. Hunting versus gathering in arid ecosystems: the evidence from the Namib Desert. In *Frontiers: Southern African Archaeology Today* (eds M. Hall *et al.*). (Cambridge Monographs in African Archaeology 10). BAR International Series 207:75–9.

Jochim, M. A. 1976. *Hunter-Gatherer Subsistence and Settlement: A Predictive Model*. New York: Academic Press.

Keeley, L. 1982. Hafting and retooling: Effects on the archaeological record. *American Antiquity* 47:798–809.

Killick, D. J. B. 1978. The Afro-alpine region. In *Biogeography and Ecology of Southern Africa* (ed M. J. A. Werger): pp. 515–60. The Hague: D. W. Junk Publishers.

Klein, R. G. 1974. Environment and subsistence of prehistoric man in the southern Cape Province, South Africa. *World Archaeol.* 5:249–84.

_____. 1978a. A preliminary report on the larger mammals from the Boomplaas Stone Age cave site, Cango Valley, Oudtshoorn District, South Africa. *S. A. A. B.* 33:66–75.

_____. 1978b. Stone Age predation on large African bovids. *J. Archaeol. Sci.* 5:195–217.

Lee, R. B. 1968. What hunters do for a living or, how to make out on scarce resources. In *Man the Hunter* (ed R. B. Lee and I. DeVore): pp. 30–48. Chicago: Aldine.

_____. 1979. *The !Kung San: Men, Women and Work in a Foraging Society*. Cambridge: C. U. P.

_____ and DeVore, I. (eds). 1968. *Man the Hunter*. Chicago: Aldine.

Manabe, S. and Hahn, D. G. 1977. Simulation of the tropical climate of the ice age. *J. Geophys. Res.* 82:3889–991.

Marks, A. 1983. The Middle to Upper Palaeolithic transition in the Levant. *Advances in World Archaeology* 2:51–98.

Martin, L. 1972. Variations du niveau de la mer et du climat en Côte d'Ivoire depuis 25,000 ans. *Cah. ORSTOM Sér. Géol.* 4:92–103.

Meehan, S. 1977. Man does not live by calories alone: The role of shellfish in a coastal cuisine. In *Sunda and Sahul* (eds J. Allen, J. Golson and R. Jones): pp. 493–531. New York: Academic Press.

Morlan, R. E. and Cinq-Mars, J. 1982. Ancient Beringians: human occupation in the Late Pleistocene of Alaska and the Yukon Territory. In *Palaeoecology of Beringia* (eds D. M. Hopkins et al.): pp. 353–81. New York: Academic Press.

Netting, R. 1971. The ecological approach in cultural study. *Addison-Webley Modular Publications* 6:1–30.

Opperman, H. 1978. Excavations in the Buffelskloof rock shelter near Calitzdorp, southern Cape. *S. A. A. B.* 33:18–34.

_____. 1984. *'n Argeologiese ondersoek van noordoos-Kaapland: Die bestaansisteem van jagter-versamelaars uit die Latere Steentydperk.* (Unpublished D. Phil. thesis, University of Stellenbosch).

Parkington, J. E. 1977. *Follow the San.* (Unpublished Ph. D. thesis, University of Cambridge).

_____. 1980. Time and place: Some observations on spatial and temporal patterning in the Later Stone Age sequence in southern Africa. *S. A. A. B.* 35:73–3.

_____. 1984. Changing views of the Later Stone Age of South Africa. *Advances in World Archaeology* 3:89–142.

Phillipson, D. W. 1976. *The Prehistory of Eastern Zambia.* Nairobi: British Institute in East Africa.

_____. 1979. *The Later Prehistory of Eastern and Southern Africa.* London: Heinemann.

_____. 1984. *African Archaeology.* Cambridge: C. U. P.

Plug, I. 1981. Some research results on the late Pleistocene and early Holocene deposits of Bushman Rock Shelter, eastern Transvaal. *S. A. A. B.* 36:14–21.

Prell, W. L., Hutson, W. H. and Williams, D. F. 1979. The subtropical convergence and late Quaternary circulation in the southern Indian Ocean. *Marine Micropalaeontology* 4:225–34.

Price-Williams, D. 1981. A preliminary report on recent excavations of Middle and Late Stone Age levels at Sibebe Shelter, north-west Swaziland. *S. A. A. B.* 37:22–8.

____ and Prior, J. 1985. Archaeological charcoals from a Swaziland rock-shelter analysed by electron microscopy as indicators of Holocene and terminal Pleistocene climatic change. (Unpublished paper delivered at the South African Society for Quaternary Research, Stellenbosch, April).

Sampson, C. G. 1970. The Smithfield Industrial Complex: Further field results. *Mem. Nat. Mus. Bloemfontein* 6:1–283.

____. 1974. *The Stone Age Archaeology of Southern Africa*. New York: Academic Press.

____. 1985. Atlas of Stone Age settlement in the central and upper Seacow valley. *Mem. Nat. Mus. Bloemfontein* 18:1-110.

Sarnthein, M. 1979. Sand deserts during glacial maximum and climatic optimum. *Nature* 272:43–6.

Schweitzer, F. R. and Wilson, M. L. 1982. Byneskranskop 1: A late Quaternary living site in the southern Cape Province, South Africa. *Ann. S. Afr. Mus.* 88:1–203.

Scott, L. 1982. A late Quaternary pollen record from the Transvaal bushveld. *Quaternary Research* 17:339–70.

Shackleton, N. J. and Opdyke, N. D. 1973. Oxygen isotope and palaeomagnetic stratigraphy of equatorial Pacific core V28-238: oxygen-isotope temperatures and ice volume on a 1,000,000 year scale. *Quaternary Research* 3:39–55.

Silberbauer, G. 1981a. *Hunter and Habitat in the Central Kalahari*. Cambridge: C. U. P.

____. 1981b. Hunter gatherers of the central Kalahari. In *Omnivorous Primates: Gathering and Hunting in Human Evolution* (eds R. S. Harding and G. Teleki): pp. 455–98. New York: Columbia University Press.

Singer, R. R. and Wymer, J. J. 1982. *Klasies River Mouth*. Chicago: Chicago University Press.

Staples, R. and Hudson, W. K. 1938. *An Ecological Survey of the Mountain Area of Basutoland*. London: Crown Agents.

Tainton, N. 1981. Introduction to the concepts of development, production, and stability of plant communities. In *Veld and Pasture Management in South Africa* (ed N. Tainton): pp. 1–56. Pietermaritzburg: Shuter and Shooter.

Talma, A. S., Vogel, J. C. and Partridge, T. C. 1974. Isotopic contents of some Transvaal speleothems and their palaeoclimatic significance. *S. Afr. J. Sci.* 70:135–40.

Tanaka, J. 1980. *The San Hunter-Gatherers of the Kalahari—A Study in Ecological Anthropology.* Tokyo: University of Tokyo Press.

Thackeray, A. I. 1981. *The Holocene cultural sequence in the northern Cape Province, South Africa.* (Unpublished Ph. D. thesis, University of Yale).

Thackeray, J. F. 1977. *Environmental change and terminal Pleistocene extinctions.* (Unpublished M. Sc. thesis, University of Cape Town).

Torrence, R. 1983. Time-budgeting and hunter-gatherer technology. In *Hunter-Gatherer Economy in Prehistory: A European Perspective* (ed G. N. Bailey). Cambridge: C. U. P.

Van Zinderen Bakker, E. M. 1976. *The Evolution of Late Quaternary Palaeoclimates.*

Volman, T. P. 1981. *The Midddle Stone Age in the southern Cape.* (Unpublished Ph. D. thesis, University of Chicago).

Webster, D. 1981. Late Pleistocene extinctions and human predation: A critical overview. In *Omnivorous Primates: Gathering and Hunting in Human Evolution* (eds R. S. Harding and G. Teleki): pp. 556–94. New York: Columbia University Press.

Wellington, J. H. 1955. *Southern Africa: A Geographical Study.* Camridge: C. U. P.

Wendt, B. 1976. Art mobilier from Apollo 11 Cave, South West Africa: Africa's oldest dated works of art. *S. A. A. B.* 31:5–11.

Whittaker, R. H. 1970. *Communities and Ecosystems.* New York: Macmillan.

Williams, J. D., Bary, R. G. and Washington, W. M. 1974. Simulation of the atmospheric circulation using the NCAR global general circulation model with Ice Age boundary conditions. *J. Appl. Meteorol.* 13:305–17.

Yellen, J. E. 1977. Longterm hunter-gatherer adaptation to desert environments: A biogeographical perspective. *World Archaeol.* 8:262–74.

Yi, S. and Clark, G. 1985. The "Dyuktai" culture and New World origins. *Current Anthropology* 26:1–20.

Human Adaptation in Southern Africa During the LGM

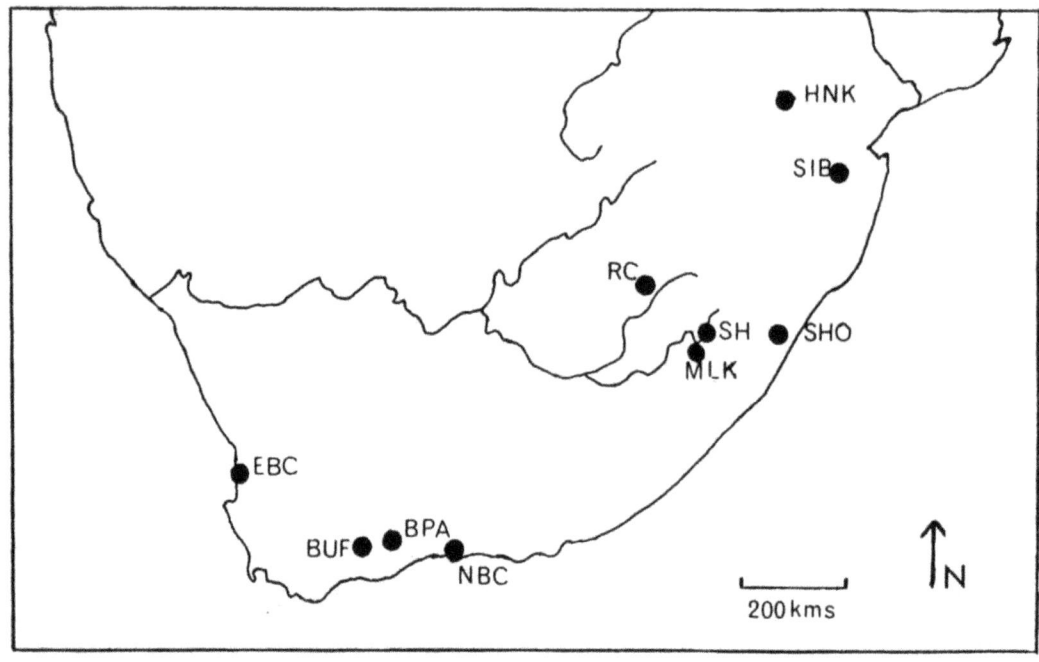

Fig. 10.1: Distribution of archaeological sites dated to the Upper Pleniglacial 25,000–16,000 bp. (Site names are given in full in Table 10.4.).

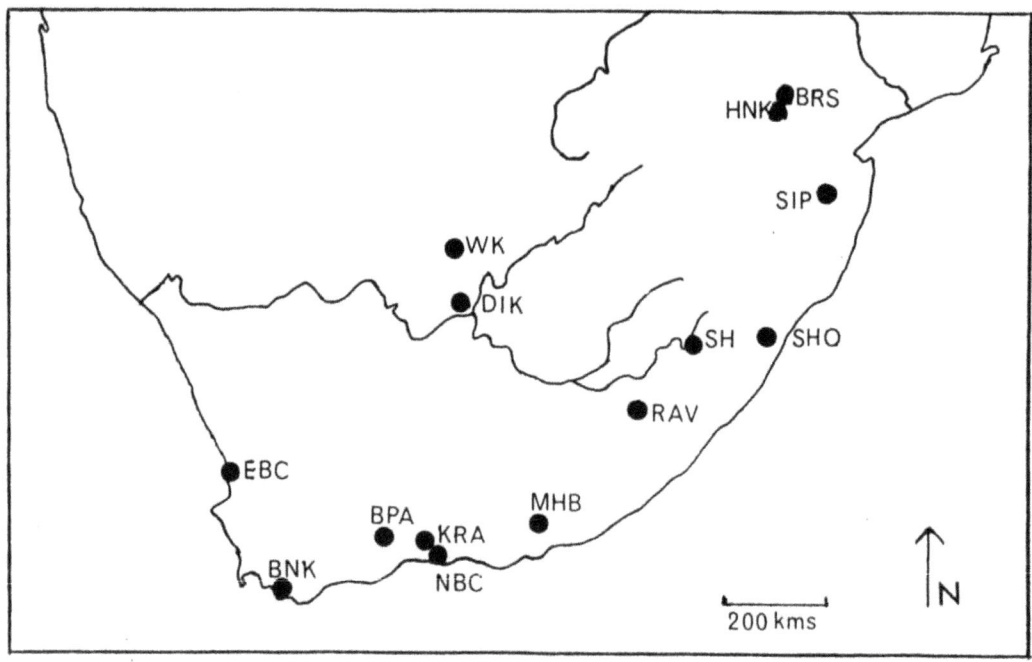

Fig. 10.2: Distribution of archaeological sites dated to the Late Glacial 16,000–12,000 bp. (Site names are given in full in Table 10.4.).

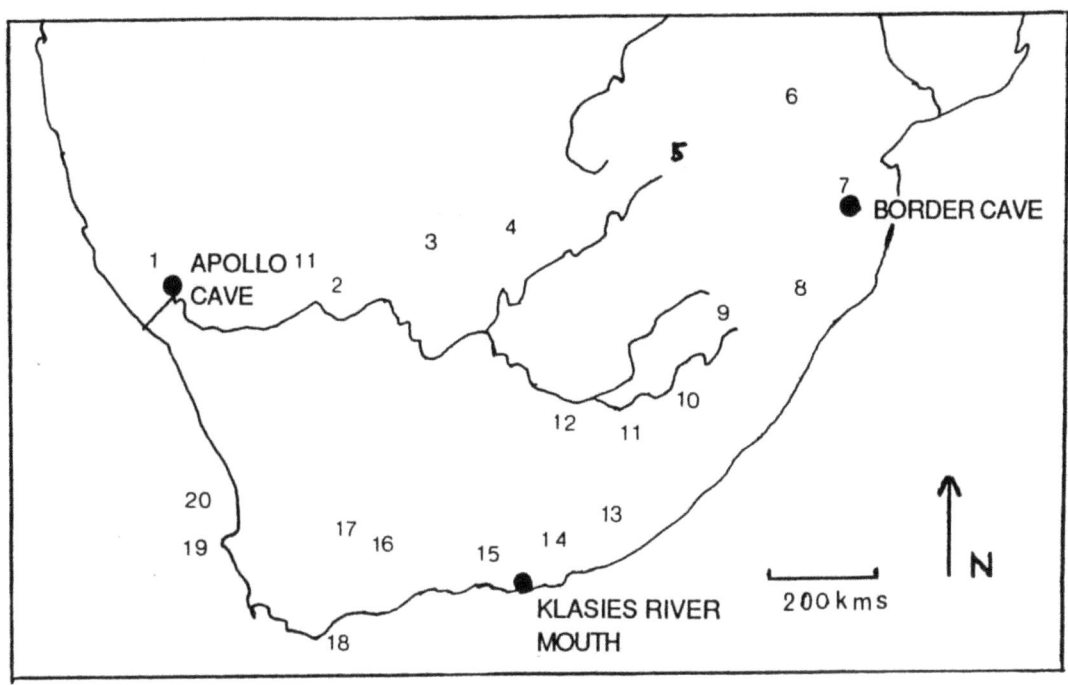

Fig. 10.3: Distribution of recent LSA research (after Parkington 1984).

1. Wendt
2. Smith
3. Beaumont, Thackeray and Thackeray
4. Humphreys
5. Wadley
6. Beaumont and Plug
7. Price-Williams
8. Mazel
9. Carter
10. Parkington and Poggenpoel
11. Opperman
12. Sampson
13. Robertshaw and S. Hall
14. Deacon, Deacon and Brooker
15. Deacon, Deacon, Inskeep and Klein
16. Deacon, Deacon and Brooker
17. Opperman
18. Avery, Schweitzer, Wilson and M. Hall
19. Smith
20. Parkington and Poggenpoel

Human Adaptation in Southern Africa During the LGM

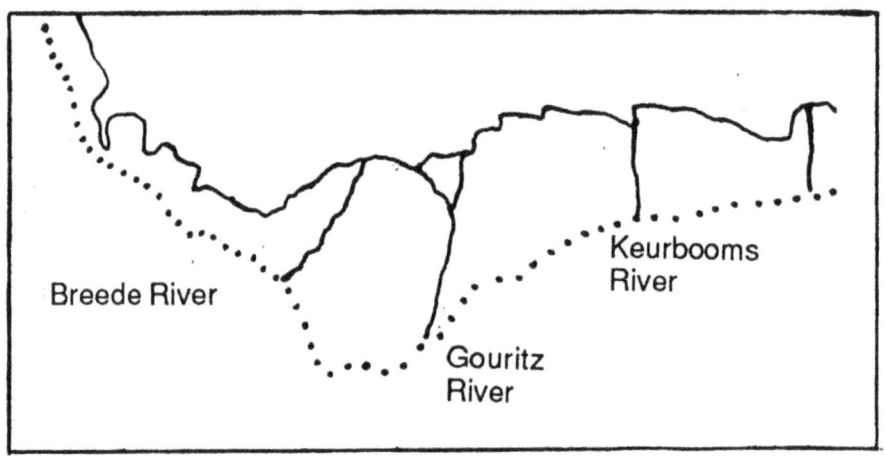

Fig. 10.4: Approximate position of the coast with a 140 m depression in mean sea level at *ca.* 20,000 bp (after Dingle and Rogers 1972).

P. Mitchell

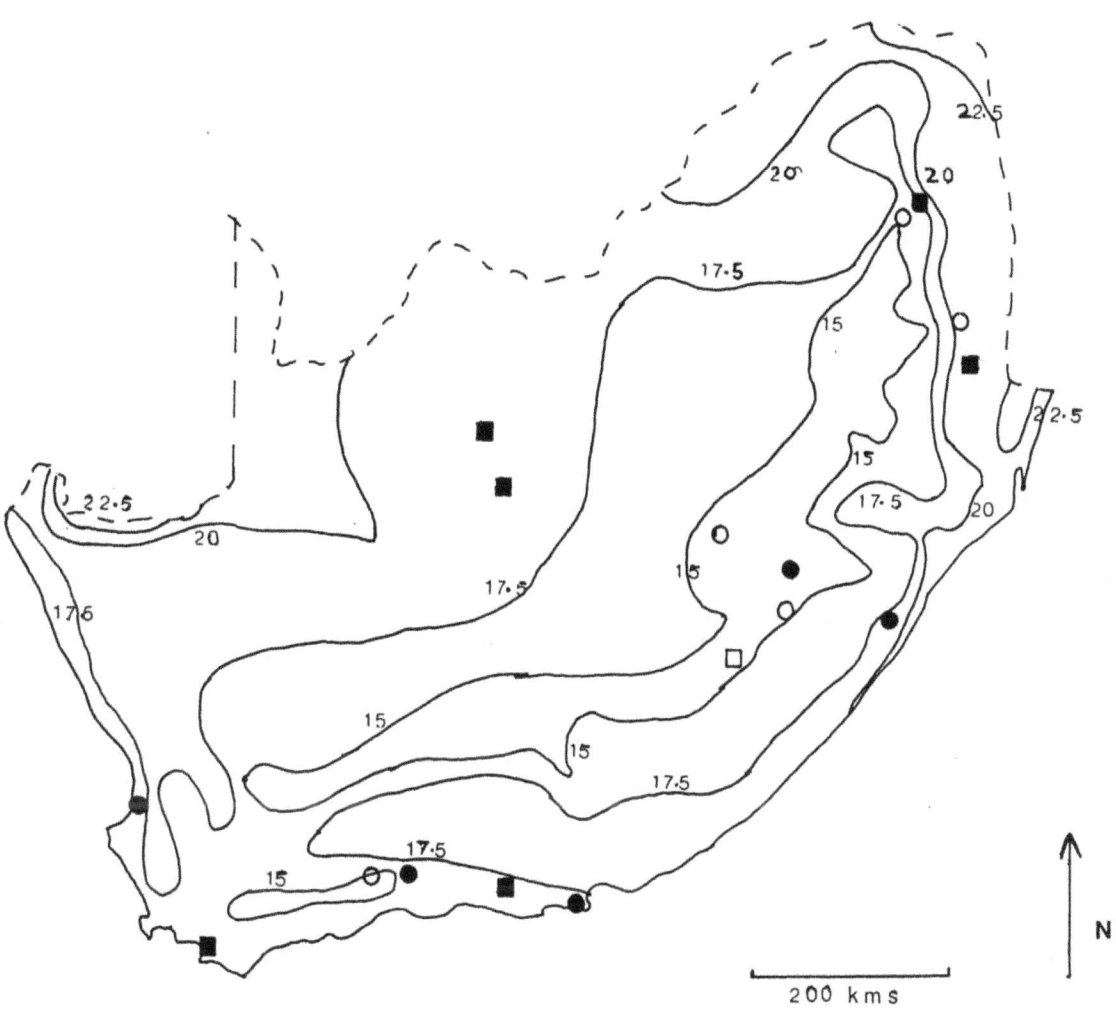

Fig. 10.5: Distribution of archaeological sites dated to the Upper Pleniglacial and the Late Glacial in relation to present mean annual surface temperature (°C).

Upper Pleniglacial site (25,000–16,000 bp) ○
Late Glacial site (16,000–12,000 bp) ■
Upper Pleniglacial and Late Glacial site ●

Human Adaptation in Southern Africa During the LGM

Fig. 10.6: Distribution of archaeological sites dated to the Upper Pleniglacial and the Late Glacial in relation to present mean annual precipitation.

Upper Pleniglacial site (25,000–16,000 bp) ○
Late Glacial site (16,000–12,000 bp) ■
Upper Pleniglacial and Late Glacial site ●

The Pleistocene/Holocene Transition in the Western Cape, South Africa: Observations from Verlorenvlei

By John Parkington, Spatial Archaeology Research Unit, Department of Archaeology, University of Cape Town, Rondebosch, Cape.

Scope

Reconstructions of terminal Pleistocene and Holocene climates along with contemporary settlement systems in the Cape are of interest for several reasons. Foremost among these, perhaps, is the fact that hunting and gathering as the prime source of food getting persisted through substantial changes in climate and geography and was replaced only in the last two millennia, apparently the result of penetration or diffusion from the north (for comment on the nature of this process see Deacon *et al.* 1978; Klein 1984; Parkington 1984). Because much of the Cape is currently only sparsely populated, a wide range of archaeological materials including substantial shell middens, rock paintings, and shelters with variable amounts of deposit survives to be used to reconstruct patterns of prehistoric behaviour. It has also been established (see Deacon 1979) that archaeological and other depositional sequences preserve valuable assemblages of plant and animal remains in sedimentary contexts offering great potential for paleoenvironmental reconstruction. It would thus seem that the Cape can act as a laboratory for studying the relationships between people and ancient landscapes, the role of climate change in determining or influencing subsistence change, and the impact of pastoralism and agriculture on hunter-gatherer communities. The juxtaposition of rock paintings with clear symbolic and ideological references (Lewis Williams 1981) and occupation sites with obvious subsistence and settlement implications (Manhire *et al.* 1984; Buchanan *et al.* 1984; Parkington *et al.* 1987 in press) allows us to see prehistoric behaviour in both its social and its ecological contexts.

Because this potential is as yet largely unrealised, we have to approach the problem of the Pleistocene/Holocene transition somewhat circumspectly. In this chapter I view the problem from the vantage point of one of only a few sequences to have been sampled from the relevant time zone, Elands Bay Cave near the mouth of the Verlorenvlei (Fig. 11.1). It is thus a single-site perspective but one sensitive to the need for a regional and spatial framework. The fact is, as we have argued elsewhere (Parkington *et al.* 1987 in press) that the paucity of terminal Pleistocene observations in the western Cape is only partly a sampling artefact and in part results from a settlement arrangement in those millennia quite distinct from that which followed in the Holocene. Thus one technique we use is the comparison of locations, sizes, distributions, and contents of sites at different time periods to emphasise aspects of change.

In order to focus on those changes which apparently relate to the global phenomenon of terminal Pleistocene climate change, but partly because of the particular stratigraphy involved, I use assemblages dated between about 16,000 and 8,000 years ago. My suggestion is that during this time there are a number of seemingly linked changes in the subsistence systems sampled at Elands Bay Cave and that the changes must at least in part be seen as responses to changes occurring in the contemporary landscape. Much of the patterning is consistent with the anticipation of H. J. Deacon published some time ago (Deacon 1976).

Paleogeography

As yet there is little primary evidence for climatic change along the western Cape coast. I use here models for such change derived from global or regional scenarios and suggest a trajectory of both climate and landscape change since the Last Glacial Maximum (LGM) of 18,000 bp. It is now generally accepted (H. J. Deacon 1983; J. Deacon *et al.* 1984; van Zinderen Bakker 1982) that the LGM conditions along the Atlantic shoreline were marked by much stronger westerly wind circulation and temperatures colder by about 5° C. The strengthening of wind systems was caused by a steepening of the thermal temperature gradient between equator and poles, but the specific effects of this at the latitude of Elands Bay Cave are as yet only partly understood. Under modern circulation patterns the region receives winter rains from the cyclonic depressions or fronts which pass across the southwestern Cape from the south Atlantic. In summer the cells are shifted some 4° south and rarely clip the Cape. One possibility is that at the LGM the area of the winter rainfall system was extended north and that more rain was experienced in summer than is at present. On the other hand, as H. J. Deacon notes (1983:326), ocean waters at 18,000 bp were cooler and thus the fronts may have been less moist at that time. A programme of coring in the Verlorenvlei is projected to test the hypothesis that local climate at the time of the LGM was colder, wetter, and windier than at present.

What is much less controversial is the suggestion that sea level was lower, by at least 100 m, and thus that the shoreline was some 35–40 km to the west of the site at 18,000 bp. Although bathymetric data are somewhat sparse, it is likely that a broad and fairly flat coastal plain was exposed. Bedrock was probably an undulating Malmesbury Shale with a partial covering of Cenozoic sands (Rogers 1987 in press). Klein and Cruz-Uribe (1987) have suggested that the range of animals in the Late Pleistocene levels at Elands Bay Cave hint at a vegetation mosaic with somewhat more grass than is present today. Currently, the area is suitable for browsers but the numbers of horses, suids, and alcelaphines suggests more graze before about 11,000 years ago.

A number of processes have acted on the landscape to change it considerably by the "post glacial" climatic optimum, taken in the southern hemisphere to fall between about 9,000 or 10,000 years ago and 6,000 years ago (Salinger 1981). The rise of sea level inundated the coastal plain, removed many thousands of km^2 from use west of Elands Bay Cave, and flooded the lower ends of such west-flowing rivers as the Verlorenvlei. Assuming that the bulk of the sea level rise took place between 13,000 and 8000 years ago, we can calculate a rough annual gain of about 2 cm vertically and 8 m horizontally. Such rates, especially when emphasised by local topographic variability, are likely to have resulted in preclimax intertidal ecosystems with pioneering species dominating the intertidal communities. Island formation and inundation is also likely to have been a regular feature of the transgressive coastline.

The flooding of the lower reaches of the Verlorenvlei will have encouraged siltation of the estuarine basin and initiated a long-term sequence of change in the riverine ecosystem. Elands Bay Cave, obviously a fixed point in the changing landscape, would have experienced extreme changes in catchment potential as the basin was drowned and filled with sediment. Our analyses of faunal material from a number of sites around the modern Verlorenvlei suggests that estuarine, lagoonal, and coastal lake conditions succeeded one another.

Along with changes in coastline morphology and estuarine ecology, we suspect there were marked changes in the local terrestrial ecosystem. Some animals such as the giant Cape horse, *Equus capensis*, and the giant buffalo, *Pelorovis antiquus*, became extinct, others became regionally less common, while animals suited to the shrub and heath vegetation began to dominate the local fauna. This may in part reflect a decrease in grazing as the shift to warmer, drier conditions took effect. The early Holocene situation is poorly represented in stratified sequences but may have witnessed an increase in hot cloudless days, a decrease in the number of effective fronts per annum, an increase in coastal fogs and a less reliable supply of fresh water to the lower reaches of the vlei. Excavations in a number of sites have convinced us that there was little use by people of the local ecosystem for about 4000 years after 7900 bp.

Archaeological Traces

At the base of the Elands Bay Cave sequence is a stony lag deposit with abundant Middle Stone Age artefacts (Butzer 1979). Above this is a sequence of ashy loams which began to accumulate before 20,000 bp and are capped by a series of shell middens. The earliest lenses of shells appear at about 11,000 years ago, and shellfish remains become the dominant sedimentary element about 1000 years later. Soon after 7900 years ago all deposition in the site seems to have ceased for about 4000 years. The uppermost deposits are shell middens with dates between 3800 and 300 bp.

In measuring changes in artefactual and faunal content, we need, of course, some stable currency, a convenient yardstick. It is normal to keep a check on the volume in cubic metres of deposit removed from each stratigraphic unit and this can serve as such a measure. Thus we know the frequency per cubic metre of faunal fragments, stone tools and ostrich eggshell pieces. Unfortunately, the rate of deposition, as also the relative contributions of natural and cultural deposition, have varied over time. The most significant such change is the increasing inclusion of shellfish in the deposits since about 11,000 years ago, obviously a change leading to much more rapid depositional rates. I have estimated, somewhat arbitrarily, that the change in depositional rate may have been more than two orders of magnitude, which even conservatively means that numbers of cultural items per cubic metre of deposit should drop very dramatically in the shell middens, *assuming no associated changes in human behaviour.*

In Table 11.1, I have shown the relative volumes of stratigraphic levels at Elands Bay Cave (using the volume of level 15 as a unit of measurement) along with their age and some aspects of their faunal and artefactual content. It is very clear from these figures that all of the items, far from declining in frequency, actually increase in the levels which see the introduction and domination of shellfish remains. As a starting point, then, we must note that fragments of animal bone and ostrich eggshell were introduced into the cave in significantly larger numbers at the terminal Pleistocene. In fact, the shift is even more dramatic when the very large numbers of marine bird, fish, and crayfish remains from levels 13 and 12 are added to the list. These could, however, be seen as "simply" a reflection of the rise in sea level and the movement of the shoreline to within exploitable distance of the cave.

The change in the frequency of stone tools seems less dramatic. There is indeed evidence for considerable continuity from the late Pleistocene loams through to the shell middens, in that the preponderance of quartz, the relative rarity of formally retouched pieces, and the high frequency of small bipolar cores continue. Throughout the sequence the most characteristic type of tool remains what we call "a naturally backed knife" made on a side-struck flake of indurated shale. What does correspond with the appearance of shellfish is an increase in the use of shale at the site and an increase in the frequency of utilised flakes. More

J. Parkington

significantly, perhaps, grindstones, whetstones, palettes, and rubbers, each a reflection of some on-site manufacturing process, all appear in level 15, admittedly in fairly small numbers. Ochre becomes dramatically more common at this time (Table 11.2).

Table 11.1: Changes in the frequencies of some items in the Elands Bay Cave sequence. Shell lenses first appear in level 15 and become dominant in level 13. There is no level 14. In translating from frequency per unit volume (/m³) to frequency per unit presumed time (/t) a factor of 10 has been applied to level 15, a factor of 100 to levels above that.

Level	Volume		OES			Mammals			Tortoise			Stone			Age
	15=1	unit	wt	wt/m³	wt/t	NISP	/m³	/t	MNI	/m3	/t	pieces	/m³	/t	Years bp
1–9	16.7	40.96	955	57.2	57	2152	129	129	127	8	8	15482	930	930	300–3800
11	3.1	7.56	394	127.1	127	552	178	178	127	41	41	285	92	92	7900–8300
12	2.4	6.00	1558	649.2	650	4867	2028	2028	570	237	237	161	67	67	8500–9600
13	2.7	6.66	5552	2056.3	2056	2554	946	946	900	333	333	198	73	73	10,000–10,700
15	1.0	2.45	647	647	65	220	220	22	217	217	22	432	432	43	±11,000
16–20	2.3	5.73	2061	896	9	1024	445	5	364	158	2	10260	4460	44	11,500–?16,000

Table 11.2: Frequency of bone and shell tools and ochre at Elands Bay Cave

Level	OES			Marine Shell		Bone Tools					ochre
	water container fragments	beads	unfinished beads	Donax scrapers	beads & pendants	awls	points	gorges	beads	dec/paint. pieces	
1–9	1	492	46	25	25	6	2	33	32	—	28
11	1	7		280	12	3	2	20	2		7
12		33	3	1	2	8	2	740	8	(7)+	10
13	5	77	10	5		3	2	4	16	3	44
15	3	17	2						49	1	25
16–20	1	56	10	1					21		10

Ostrich eggs, I assume, were eaten and the shells used as water containers and made into beads. Tables 11.1 and 11.2 make it quite clear that all of these activities leave much more visible residues at the site after 11,000 years ago, particularly if depositional rates are taken into account. Along with this go very substantial increases in bone tools, decorated bone and shell, shell tools, and, less abundant, pieces best described as byproducts or unfinished items. Much of this, too, can be seen as evidence for on-site manufacture.

There can be little doubt, I think, that we see here a strong signal of changing behaviour at this particular site. To refer to this in "cultural" terms as the Albany Industry (Deacon 1978) is to imply that the change is widespread and reflects the emergence of a new system. I prefer to see these patterns as, at least in part, a reflection of change within the system, reorganisation perhaps in the context of a changing paleogeography. In making this case I argue that there is an underlying continuity in the character of the stone tool assemblage and in some components of the faunal assemblage. What seems to change are 1) the frequency with which material is returned to the site, 2) the visibility of arguably domestic manufacturing debris, and 3) the diversity of both faunal and artefactual assemblages. I suggest that a dominant feature is the increasingly domestic character of site visits as the shore moved closer to Elands Bay Cave.

The appearance of shellfish, marine mammals, coastal birds, fish, and crayfish at the site undoubtedly signal a change in behaviour. The fact that the frequencies of terrestrial mammals, tortoise bones, and ostrich eggshell fragments all increase at the same time mean that the change is more than simply a shift of attention to the sea shore. More subtle associated changes, such as that toward a higher frequency of newborn individuals among the small bovids (Parkington 1981), also imply organisational change. Although hard to demonstrate, it seems that many small food parcels of the kind gathered by women and children in the vicinity of a home base appear in the site soon after 11,000 years ago. Shellfish are among these. I should note, though, that poor preservation of plant materials is a serious problem in establishing this point.

The environmental context of this change, described earlier, is that of a transgressive sea level and a rapidly shrinking coastal plain. During the time period of the lowermost levels at Elands Bay Cave the site lay some 35 km inland in a very prominent cliff face adjacent to an active west-flowing river. For several millennia, perhaps as many as eight, bones and stones accumulated in the site very slowly. Very infrequently beads, ochre, and other decorated items were lost there. Coincidently with the decision to begin to exploit shellfish and other marine animals from an approaching shoreline, this pattern was transformed. Both the diversity of animals collected and the range of stone tools produced increased markedly. Perhaps related to this is the fact that we have located four burials in the millennium or so between 10,800 and 9600 years ago, but none from the preceding 10,000 years. I suggest that prior to 11,000 years ago the site was a partial and perhaps not fully representative sample of the range of debris left by contemporary prehistoric people. The partiality may have been the result of very temporary occupations, stopovers between more substantial visits elsewhere in the landscape. Or, alternatively, the location of the site may have made it an attractive place from which to pursue some particular set of activities, seasonally or from time to time. In this latter scenario we could envisage brief and quite specific occupations by some limited segment of the community, thus accounting for the very low diversity in debris when compared with later levels. By the time sea level was approaching that of today the home base had been shifted eastward to appear by 11,000 years ago in Elands Bay Cave.

All of this is not to deny that the system, of which parts consistently were deposited in the cave, was changing. I merely suggest that much of what appears in the site along with shellfish may already have existed and been left in sites now unfortunately covered by the rising sea. What does seem to be unavoidable is the recognition of a substantial role for the changing environment in bringing about changes in prehistoric human behaviour. I suggest that to avoid the charge of environmental determinism we use the concept of place (Parkington 1980; Binford 1981) as a link between environments and people. Obviously an environment or a change in environment cannot change people's behaviour; they have to do that for themselves. From the perspective of Elands Bay Cave I suggest that the effect of climatic and landscape changes at the end of the Pleistocene was to transform the set of places available to prehistoric hunter-gatherers. People, thus, perceived the changing attractiveness of specific locations and reorganised their behaviour to suit. Places are "spaces given meaning" and Elands Bay Cave simply became a different place.

Acknowledgements

Field work at Elands Bay Cave has been supported by the University of Cape Town, the Human Sciences Research Council, the Council for Scientific and Industrial Research and the Swan Fund.

References

Binford, L. R. 1982. The archaeology of place. *J. Anthrop. Archaeology* 1 (1):5–31.

Buchanan, W. F., Parkington, J. E., Robey, T. S. and Vogel, J. C. 1984. Shellfish, subsistence and settlement: some Western Cape Holocene observations. In *Frontiers: Southern African Archaeology Today* (eds M. Hall, G. Avery. D. M. Avery, M. L. Wilson and A. J. B. Humphreys): pp. 121–30. (B.A.R. Int. Series 207).

Butzer, K. W. 1979. Geomorphology and geo-archaeology at Elandsbaai, Western Cape, South Africa. *Catena* 6:157–66.

Deacon, H. J. 1976. Where hunters gathered. *S. A. Arch. Soc. Monograph Series* 1: 1–232.

———. 1979. Excavations at Boomplaas Cave—a sequence through the Upper Pleistocene and Holocene in South Africa. *World Archaeology* 10(3): 241–57.

———. 1983. Another look at the Pleistocene climates of South Africa. *S. A. J. Sci.* 79:325–8.

———., Deacon, J. Brooker, M. and Wilson, M. L. 1978. The evidence for herding at Boomplaas Cave in the Southern Cape, South Africa. *S. Afr. Archaeol. Bull.* 33:39–65.

Deacon, J. 1978. Changing patterns in the late Pleistocene/early Holocene prehistory of southern Africa as seen from the Nelson Bay Cave stone artefact sequence. *Quatern. Res.* 10:84–111.

———., Lancaster, N. and Scott, L. 1984. Evidence for Late Quaternary climatic change in southern Africa. In *Late Cainozoic Palaeoclimates of the Southern Hemisphere* (ed J. C. Vogel). Rotterdam: A. A. Balkema.

Klein, R. G. 1984. The prehistory of Stone Age herders in South Africa. In *From Hunters to Farmers* (eds J. D. Clark and S. A. Brandt): pp. 281–90. Berkeley: University of California Press.

———. and Cruz-Uribe, K. 1987. Large mammal and tortoise bones from Elands Bay Cave and nearby sites, western Cape Province, South Africa. In *People and Places: Papers in the Prehistory of the Western Cape* (eds J. Parkington and M. Hall). Oxford: B. A. R. Series.

Lewis-Williams, J. D. 1981. *Believing and Seeing*. New York: Academic Press.

Manhire, A. H., Parkington, J. E. and Robey, T. S. 1984. Stone tools and Sandveld settlement. In *Frontiers: Southern African Archaeology Today* (eds M. Hall, G. Avery, D. M. Avery, M. L. Wilson and A. J. B. Humphreys): pp. 111-20. Oxford: B. A. R. Int. Series 207.

Parkington, J. E. 1980. Late Pleistocene and Holocene climates as viewed from Verlore vlei. *Pal. Africana* 23:71.

———. 1981. The effects of environmental change on the scheduling of visits to the Elands Bay Cave, Cape Province, S.A. In *Patterns of the Past* (eds I. Hodder, G. Isaac and N. Hammond). Cambridge: Cambridge University Press.

_____. 1984. Soaqua and Bushmen: hunters and robbers. In *Past and Present in Hunter Gatherer Studies* (ed C. Schrire). Orlando: Academic Press.

_____. *et al.* 1987. Holocene Coastal settlement patterns in the western Cape. In *The Archaeology of Prehistoric Coastlines* (eds G. Bailey and J. Parkington). Cambridge: Cambridge University Press.

Rogers, J. 1987. The evolution of the western Cape continental terrace between St. Helena Bay and Lamberts Bay. In *People and Places: the Prehistory of the Western Cape* (eds J. Parkington and M. Hall). Oxford: B. A. R. Series.

Salinger, M. J. 1981. Palaeoclimates north and south. *Nature* 291:106-7.

van Zinderen Bakker, E. M. 1982. African palaeoenvironments 18,000 yrs bp. *Palaeoecology of Africa* 15:77-9.

The Pleistocene/Holocene Transition in the Western Cape

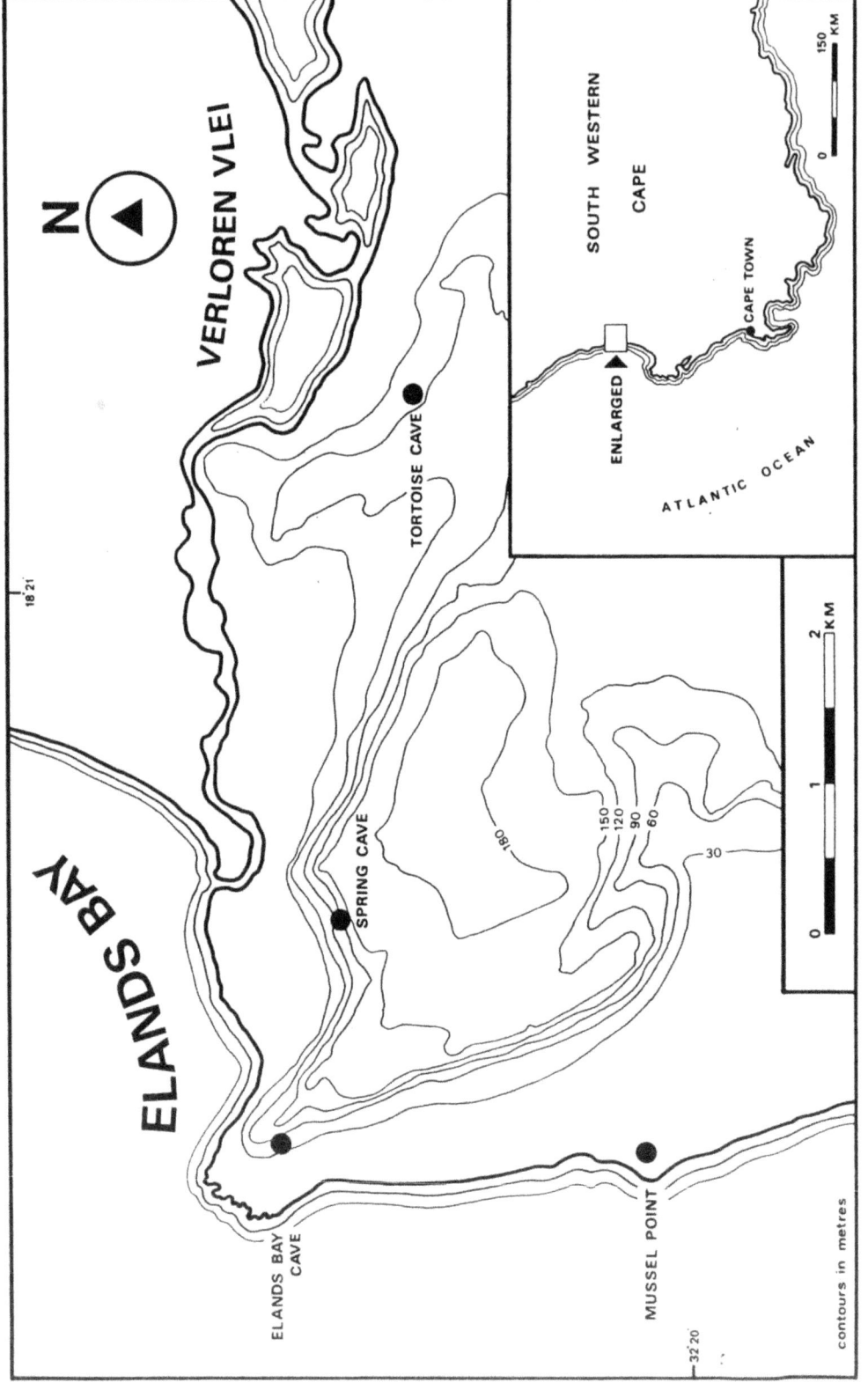

Fig. 11.1:   Map of Elands Bay.

Putting the Wind Up the Smithfield: Seasons of Occupation Inferred for Sub-Recent Bushman Surface Sites

By C. Garth Sampson, Department of Anthropology, Southern Methodist University, Dallas, Texas.

Seasonal Mobility

Hunter-forager groups who live in marginal habitats exploit seasonally abundant food sources with patchy distributions. They distribute themselves on the landscape in ways that give each group more or less exclusive access to a set of contiguous food patches. The schedule of a group's movements between its patches will be largely determined by when and where food is available. Also, the group may break up seasonally into subgroups each with temporary rights to a patch (or patches). Seasonal mobility is, therefore, the key to survival in marginal habitats. Because it is so crucial as an adaptive device, it inevitably becomes the cornerstone of any ecologically oriented analysis of hunter-foragers, present or past.

Attempts to reconstruct prehistoric models of seasonal mobility all take the above paradigm as their point of departure. A local model of seasonal mobility is fitted to a particular landscape (the selected study area) and its resources, then a settlement pattern is intuitively derived from the model. This pattern inevitably predicts that sites would have been occupied at different times of year in different parts of the area.

The archaeological implications of this prediction are that individual sites will contain food waste eaten at the one time of year only, no matter how often the site was reoccupied. Thus all such studies start by determining the season of occupation of several different sites scattered about the study area.

Both the theoretical underpinning and methodology of this approach have been extensively reviewed (Monks 1981), and only three of the weakest points need be mentioned here. Foremost is the inherent, but untested, assumption that the selected study area contains a cluster of food patches over which *one* group exercised more or less exclusive rights, and consequently had absolute freedom of movement. If, in fact, two groups had rights to separate fragments of the study area, then pieces of two different settlement patterns will be mapped. Without prior knowledge of the territorial boundary separating them, the archaeologist who is trying to "explain" the map of seasonally occupied sites will inevitably do so in terms of the movements of a single group, not of two adjacent groups.

Putting the Wind Up the Smithfield: Seasons of Occupation

The season of occupation of the individual site is reconstructed by the analysis of food waste. The standard assumption is that modern seasonal changes in the food supply surrounding the site are valid analogs of past seasonal availability. The standard analytical procedures are: 1) seek ethnohistoric accounts of seasonal exploitation as a test of the standard assumption; 2) isolate from the site's residue those remains of plants and/or animals known to be seasonal markers today, e.g., migratory birds; 3) identify those elements with incremental growth stages that reflect the season-of-death, e.g., fish scales, otoliths, molluscs, or tooth eruption sequences in young mammals. Wherever possible, independent data sets are sought from the same site. Overlapping results are taken to indicate seasonal occupation of the site, whereas conflicting results are taken to signify multiseason or even year-round occupation. Where only single tests are possible at individual sites, results should covary with the intuitive model derived from studies of the surrounding resources.

The second untested assumption now comes into play: that the presence of a seasonal indicator (i) in the food waste of a site means that the site was occupied *only* in that season. However, the same level in the same site may contain other foods (t,u,...) which do not yield seasonal information, but were in fact eaten at any or all the other seasons of the year when (i) was not available. This means that the site was in fact randomly reoccupied at many different times of year, but that the time when (i) was available is the only season that is being registered in the archaeological record. If this is so, then a map with several sites at one end of the study area (all yielding a spring indicator), and with several sites at the opposite end (all with a fall indicator) does not demonstrate that people moved back and forth across the map in metronome fashion. It is simply a map showing where those items were eaten at the times when they were available.

The third limitation of this approach is geographical: it can be applied only in areas yielding seasonal indicators. Consequently, in Africa, such studies are virtually confined to arid lands (plant remains), or coastal belts (fish, molluscs, and seabirds).

That the first two assumptions should be taken seriously has been demonstrated recently by an independent test of a finely argued model of winter/coast to summer/inland mobility for the southwestern Cape Province, South Africa (Fletermeyer 1977; Parkington 1972, 1984). The isotopic analysis of human skeletons buried on the coast indicates an exclusively coastal diet, and the smaller sample of inland skeletons suggest an exclusively montane diet (Seeley and van der Merwe 1985). The implication of this result is that free movement across the study area was restricted, and that boundaries between groups *were* maintained. Isotopic analysis offers great promise as the first truly independent cross-check of the standard approach to seasonal mobility, but will apply only where skeletal material is available. It will no doubt be used routinely in the future, together with the methods already outlined. Clearly, other lines of independent testing are urgently needed in order to build stronger cases for seasonal mobility.

Windshelter Analysis—Principles and Assumptions

The term *windshelter* is coined here to mean the shelter afforded to an archaeological site from any prevailing wind by natural rock barriers. This chapter attempts to explore the potential and limitations of windshelter as an indicator of seasonal occupation at an archaeological site. It will exploit the simplistic notion that a site sheltered from one seasonally prevailing wind will be occupied more frequently during that season, and less frequently at other times of year, when prevailing winds are from other directions and the site is exposed.

This approach is loaded with its own peculiar set of prior (untested) assumptions. They require careful review before a particular model for a selected study area is proposed. The most important assumptions are: 1) modern prevailing wind directions and velocity patterns are valid analogs of past wind patterns for the prehistoric period under consideration; 2) hunter-foragers in the chosen study area routinely took wind direction and velocity into consideration when deciding where to set up camp; 3) temporary structures/shelters built on the site were oriented according to social rules, rather than all-backs-to-the-wind. This study attempts to test only the first two assumptions. Like the standard approach, this method carries with it the weakening assumption that the selected area is not divided by territorial boundaries. Additional assumptions may be demanded by a specific model in its landscape.

Windshelter analysis offers both advantages and disadvantages over the standard methods outlined in the previous section. At the theoretical level, it may be reasoned that a site sheltered from one prevailing wind was used *exclusively* during the season of that wind, because at other seasons it would be exposed to the full force of various other winds. If it was necessary to visit the area at some other season as well (water, food plants), then another appropriately sheltered spot nearby would be chosen.

At the practical level, windshelter analysis opens up the possibility of seasonal labels for sites without seasonally sensitive organic remains. This includes surface lithic scatters with no organic residues surviving. Furthermore, there is the possibility that a seasonal label can be obtained without excavating the site. Preliminary data can be gathered during survey, at the time when the site is discovered and first recorded. Because the method is quicker, cheaper, and less demanding of specialist expertise, much larger maps can be compiled than by the normal approach.

Like the latter, however, it cannot be used everywhere. It is only really viable in those areas where wind velocities become very high at certain times of year. Furthermore, it is only really applicable in areas with relatively little tree cover, such as semi-desert and steppe country. Then again it can only be applied in such areas for which adequate meteorological data are available. These must include diurnal as well as longer-term variations, together with temperature and precipitation records. The database should also be of long enough duration to allow averaging of data, thus eliminating year-to-year noise in the wind record. In short, there must be a long-established meteorological station close enough to the study area that its record will be comparable. If this requirement cannot be met, then the cost of acquiring the data becomes prohibitive.

This approach is also not viable in areas where caves are the chief focus of settlement. The advantages offered by a deep cave are such that it is not likely to be avoided simply because its mouth is oriented into a prevailing wind (e.g., de Lumley 1969). Even in areas where rock shelters are superabundant, those chosen for occupation have not been found to be uniformly oriented (e.g., Mobley 1981), but were selected because of size, ease of climb, view, and other more subtle strategic factors. The method holds out most promise for areas where caves and shelters are rare and small.

Wind Pattern Analysis—Methods

The modern wind record must be analysed first in order to erect a predictive model of how (from what direction) and when (in what season) sites will be found to be protected. The most convenient device for averaging and summarising these data is the windrose. This portrays graphically the amount (duration and velocity) of wind blowing for a chosen time segment. Eight or sixteen vectors (usually compass points) may be selected according to need and/or the nature of the record.

The first windrose needed is the *Mean Annual Windrose*, which should be derived from at least five year's continuous record, or more if possible. Anything of shorter duration will probably contain a fair amount of short-term noise. This initial summary diagram will determine that there are, indeed, prevailing winds, and will demonstrate their direction and the extent of their directional variability. Clear directional spikes must be present; otherwise no predictive model will be forthcoming.

The second task is to construct twelve *Mean Monthly Windroses* from the same data set. These will help the archaeologist to begin isolating wind directions that have exceptionally high velocities for limited periods of the year. This chart allows the first crude predictive model to state which windshelter directions are likely to indicate which seasons of occupation. These statements may be further refined, since humans tend to react to extremes of weather, not to averages.

The third step is to analyse extremes of wind velocity and temperature on a daily basis over a complete year, chosen at random from the record. This will reveal the seasonality of short-term bursts of high gusting winds, and identify winds which tend to bring freezing weather. Such details are likely to be drowned out in the averaged windroses. Many other useful associations are likely to be revealed in this chart.

Next, the directions of the highest recorded velocities can be extracted, averaged, and plotted on a separate *High Velocity Windrose*. The monthly frequency of high-velocity winds from specific directions can then be plotted to determine if they occur seasonally. If so, then the model may acquire additional seasonal windshelter vectors to accommodate these special winds.

The fifth task is to determine the directions of night winds during winter months. Windchill factors can be computed from the data and a *Winter Night Windrose* can be constructed. This will demonstrate whether the most extreme cold periods of the yearly round are associated with specific wind directions or not. Further refinements are possible here also where sudden drops in night temperatures can be tied to shifts in wind direction.

More fine-grained plotting, perhaps even at hourly intervals, will reveal such associations. This procedure may have especially strong predictive value in areas where winter night minima are well below freezing. It may prove essential to shelter at night during winter from certain near-lethal wind directions, not because they blow regularly, but because they *might* blow on any given night.

Windshelter Analysis—Methods

The predictive model is now ready to be tested in the field. Windshelter has two potentially measurable components: direction(s) and effectiveness. The *direction* of shelter is expressed in degrees and/or compass vectors, and is measured by shooting angles from a midpoint on the site to the edges of the sheltering outcrop(s). The *effectiveness* of shelter is a complex of factors including the kind of outcrop(s), height(s), shape(s), and distance between site and object. Attempts to combine these factors produce rule-of-thumb estimates at best. Truly reliable measurements of effectiveness can only be obtained by plotting wind speed on a map of the site while the prevailing wind is blowing. A single measurement can be obtained by averaging these plots, or by some ratio of mean wind speed to site surface area. Detailed, accurate measurements will be much more time-consuming and expensive than rule-of-thumb estimates, such as a numbered ranking system of effectiveness of subvectors of windshelter arc around the edge of the site. Thus the highest sector of a protective ridge could be assigned a value of 3, while its lower shoulders or gaps would be assigned 2 or 1. These values are obviously subjective, and their repeatability from one observer to the next remains in doubt.

The first analytical task after the field data are gathered is to subdivide the sample into groups according to the number of protective vectors. Thus sites protected from N-NNW-NW would be placed in the 3-vector class. Each vector class is then plotted on its own *Windshelter Rose* and those classes with large enough samples may be tested for randomness of distribution through the 16 vectors.

If distributions are not random, and if the cause of the nonrandom distributions are spikes in the rose which correlate with spikes appearing in the preceding Wind Pattern Analysis, then assumptions 1 (present winds=past) and 2 (hunter-foragers sheltered from wind) are at least partially tested and confirmed. If they are randomly distributed, then further analysis of the existing wind patterns may be necessary to tease out the discrepancies. If this fails, then the assumptions are probably invalid and the test should be abandoned. Obviously, it is unwise to enter a project of this kind without some prior hunch that there will be a close fit between wind patterns and windshelter. This comes from detailed analysis of the meteorological data and plenty of first-hand experience of the sites and their settings, not to mention first-hand experience of the winds!

Assuming that all is well and that the analysis may proceed, the final task is to cluster sites by shared windshelter, and to plot each group on the basemap. Spatial clustering of sites within each windshelter type is then sought. Spatial contrasts in the clustering of the various types may now become apparent. Appropriate inferences about seasonal mobility are then drawn. This paper deals only with the testing of the basic assumptions on which seasonality is inferred, not with plotted results.

The Test Case—Background

The selected area for this trial study is the headwaters of the Seacow (also Seekoei, Zeekoe) River valley in the Upper Karoo of the central South African plateau (Fig. 12.1). It comprises 2065 km^2 of the upper reaches of a small tributary of the central Orange River basin, and is typical of the Karoo landscape with windswept, treeless flats interspersed with dolerite ridges and hill swarms which provide the only available windshelter. The latter have scattered low bush and scrub cover, and are associated with reliable spring eyes plus abundant outcrops of quality hornfels. Karoo scrub vegetation has a remarkably high carrying capacity, now supporting mainly sheep and a very small residue of game animals, which abounded here up to the early 1800s.

The Zeekoe Valley Archaeological Project (ZVAP) began in 1979 with an archaeological survey of the central and upper Seacow valley. Exploration was halted in 1981 when the surface archaeology of some 5000 km² had been recorded. This survey and subsequent seasons have produced over 16,000 surface sites attributed to at least eight consecutive occupation events. The stratigraphic ordering of these events was previously documented in the central Orange River valley (Sampson 1972). Site distribution maps of each episode indicate substantial changes in the settlement pattern through time (Sampson 1985:105). The first occupation is certainly of mid-Pleistocene age, probably a quarter-million years ago (Partridge and Dalbey in press), and the last occupation is referred to the Smithfield Industry (Sampson 1974:374ff). This was still the prevailing artifact tradition of the Bushman hunter-foragers when European *trekboers* arrived in the late 1700s (van der Merwe 1937). The earliest available radiocarbon dates associated with Smithfield tools and decorated pottery are from the fourteenth century (Beaumont and Vogel 1984:93). It is doubtful whether either was made much later than 1800 AD.

Over 6000 Smithfield sites have been identified. These sites can be readily separated into quarries, smaller lithic scatters, and "camps." The camps are larger, denser lithic concentrations with a wider range of artifact types, including grinding equipment and pottery. They are usually concentrated on flat sandy spaces on the footslopes and crests of dolerite hills and ridges. Most of the camps are within a 1 km radius of the permanent spring eyes (Sampson 1984). Springs were the nodes of the food patches in this habitat. Camps are nearly always carefully positioned so that the spring is visible from a position near the the camp rim, but the camp is invisible to game drinking at the spring. Certain springs were either avoided or used exclusively for stalking game. There may have been a loose system of "tandem" spring use in which one eye was for drinking water while its nearest neighbour was reserved for game, but this was not invariably the pattern.

The Upper Seacow Bushmen were obviously highly mobile on the landscape, as are modern Kalahari San (only 500–700 km to the north), whose mobility patterns present several variations (Barnard 1979). Brooks (1984) has proposed a valuable summary ethnographic analog

Putting the Wind Up the Smithfield: Seasons of Occupation

which can be reasonably adapted to the upper Seacow mobility pattern. A Bushman group would settle near a waterhole until the surrounding food plants and/or firewood ran out, and/or the game drifted away. They would then move to another waterhole, having carefully assessed its resources and considered how long its catchment had been rested since the last visit. On arrival, they seldom occupied the same camping spot used on their previous visit as this was not yet free of waste, feces, and ticks. If it was last used at a different time of year, it might have been prudent to shelter from some other wind direction. If the prevailing wind shifted during this next visit, it would be no great task to set the camp on one of several other sheltered spots on the same ridge. Thus it came about that Smithfield camps are clustered in the vicinity of certain spring eyes, and surrounded by smaller work stations up to 2–3 km from the camping area (Sampson 1984).

Studies of the decorative motifs on potsherds from some 900 camps in the upper valley show that certain motifs have clearly restricted distributions (Sampson in press). Furthermore, the distributions of two of these motifs coincide with the distribution of a rare rock type found among the flakes of certain camp residues. The edges of these overlapping distributions coincide with linear features on the landscape such as dongas or low dolerite dykes. Our current working hypothesis is that these features were the visible reference points used by the Bushmen to denote mutually agreed-upon territorial boundaries between neighbouring bands. The ceramic studies allow us to propose a spatial model containing one complete San territory and large portions of three contiguous territories within the selected area. The size of the complete example is well within the known range of territories exploited by living Kalahari San (e.g., Hitchcock 1982:249), but it contains many more permanent waterholes than any known Kalahari territory.

A Predictive Model Derived from Modern Wind Patterns

Hourly weather readings are taken at the Grootfontein College of Agriculture and Research Institute located only 29 km due east of the study area (Fig. 12.1), which is close enough to accurately reflect wind patterns and other data in the upper Seacow drainage. Only limited data from this station have been published (SA. Dept. of Transport 1960:34–5).

An overview of wind variations is shown in the Mean Annual Windrose in Fig. 12.3. This serves to demonstrate what every local resident knows: the Karoo is a very windy region. It also shows that there are two prominent wind directions, broadly described as northwesterlies (actually N-WSW) and southeasterlies (actually SE-E). If only quantity (duration and velocity) are considered, the NNW-W winds contribute more than SE-E winds. Other directions contribute minimal amounts. This diagram provides the basis for a predictive model: if assumptions 1 and 2 are valid, then Smithfield camps should be protected from these two general directions.

Average Monthly Windroses are given in Fig. 12.4. There is no direction which occurs exclusively in one season, but durations of wind for single directions vary with the seasons. NNW-W winds start to build in mid-winter (June) and peak in July/August. Extreme NW gusts occur in September. The progress of the northers sees an increase in duration (but not velocity) which peaks in midwinter. Summer sees an increase in the duration (but not velocity) of SE-E winds. Based on these data, we tentatively add to the model's predictions: sites sheltered from NNW-W were used more in late winter/early spring; N-sheltered sites were used mainly in midwinter; sites sheltered from SE-E were used more in summer. These will be modified as the analysis proceeds.

Fig. 12.5 tracks the maximum velocity and maximum gust for every day of a randomly chosen year (1978) from the unpublished hourly records. Velocities fluctuate in quasi-cyclic fashion, as do the temperature extremes. High winds are associated with higher temperatures, and most of these episodes are from NNW-W. Days with light winds tend to be cooler, and are mainly from the SE-E vectors. [For a climatological explanation of these relationships, see Tyson (1969).] Thus it emerges that wind direction changes completely *every few days*. Obviously, the model in its present form is far too simplistic. The SE-E winds are now seen to be beneficial (cooling) in summer, so that sites sheltered from these directions are poor candidates for summer sites. In winter, however, these same winds co-occur with subzero temperatures. SE-E protected sites should be reassigned as winter sites. At this stage, the model now predicts: *NNW-W=late winter/early spring; N=winter; SE-E=winter.*

Fig. 12.6 shows the direction of high velocity winds only. For most of the time, these are blowing from only four vectors. These northwesterlies reach a peak in July-August, but are actually more noticeable in September because of tremendous gusts. These occur usually in the midafternoon with much dust, and with enough force to knock a person to the ground. Walking under these conditions becomes exhausting after a few hours. Although July-August are windy months, there are other more discomforting (colder) winds to be avoided in winter. Unless camp was shifted every few days to accommodate wind changes, it would make more sense to move to NNW-W protected sites in September as the weather warms up. Thus the model is further adjusted to predict: *NNW-W=spring; SE-E=winter; N=winter.*

Putting the Wind Up the Smithfield: Seasons of Occupation

An important feature of winter winds is the marked contrast in average night and daytime direction of airflow (Fig. 12.8). Almost all night wind blows from N-NNW, in marked contrast to the more diffuse windrose for daytime. Northern airflow at night has a very clear winter seasonality (Fig. 12.9). Windchill from winter northers makes shelter from this direction essential on freezing nights. Once again, we may modify the model to read: *NW-W=spring; SE-E=winter; N-NNW=winter nights.*

The association between SE-E winds and cold conditions in winter require closer examination. Fig. 12.4 and 12.8 show that these winds contribute a trivial amount to winter wind. When "cold" days (maximum temperature below 10° C) are selected from a 10-year record, they average only 9.25 days per year. Wind directions during those days are dominantly from SE-E (Fig. 10) and cold days are, obviously, a winter phenomenon (Fig. 12.11). The windchill induced by these winds makes it imperative to seek shelter; thus the model can now be altered to: *NW-W=spring; SE-ESE=(rare) winter days; NNW-N=(most) winter nights.*

Another notable feature of the 10-year sample of "cold" days is that the temperature drops suddenly whenever the wind veers to true south. This happens on average only 2.2 days per year, but the effects are memorable. Two typical examples are shown in Fig. 12.12 and 12.13. Either the daytime maximum temperature is held down as the result of constant southerlies (Fig. 12.12 *left*), or the normal morning temperature rise will reverse abruptly (Fig. 12.13 *left*), as the wind veers to south. Both examples are contrasted with normal hourly patterns for typical winter days when northwesterlies are blowing. Such events are extremely rare at night—the analysis of June nights over five years (Fig. 12.8) revealed only four occasions (12 hours total) when winds veered to true south. However, they are so devastating that wildlife has been known to die, and stock herds are threatened. Of all winter wind directions, true south appears to be the most lethal. To the model may now be added: *S=(very rare) winter nights.*

Finally, mention should be made of summer conditions. Being farther south, Karoo summer highs seldom, if ever, reach 35° C, in contrast to the well-known extremes suffered by modern Kalahari Bushmen (who do not have to endure such bitter winters). As noted, hot windy days are interspersed with milder breezy days. Unfortunately, we do not yet have hourly temperature records from a sheltered Smithfield camp, but subjective experience suggests that these often rise well above 35° C on calm summer afternoons. The reason for this anomaly is that the black dolerite ridge, cliff, or boulder which shelters the camp on windy days is now radiating, and drives the local ambient air temperature to insufferable heights. Calm days are the worst, as shown in the examples in Fig. 12.14. It would be prudent, therefore, to locate in an exposed position on a ridge crest or, if there are electrical storms about, in some other exposed place in order to catch whatever cooling breeze is available, and to gain relief from flies. Thus, the final addition to the model is: *exposed=(mostly) summer.*

The completed inventory of predictions in the model can now be tabulated anti-clockwise:

Windshelter needed from	Season
N-NNW	(most) winter nights
NW-W	(mostly) spring
S	(very rare) winter
SE-ESE	(rare) winter days
Exposed	(mostly) summer

Southwesterlies and northeasterlies play no significant part in the model. Also, the model provides for no specific windshelter direction in the fall.

From the viewpoint of the Bushmen, it would be tiresome to move camp every few days or (in winter) every morning and evening just to accommodate shifts in wind direction. However, such moves *could* be made around ridges and hills at individual waterholes, where a wide variety of sheltered spots is usually available a few hundred metres apart. No doubt game and food plant supplies dictated the timing of between-waterhole moves, but it is intuitively reasonable to assume that wind shifts caused short-term moves between the camp sites while the group was in residence at a given waterpoint. Another likely cause of short-term moves would be the disposition of game herds on the surrounding plains, as it would be important that residents of the camp be able to observe game movements without themselves being too prominently visible. On warm days, it might even pay to camp in the teeth of a northwesterly, just to be positioned upwind from a nearby herd. Thus the choice of camp site during any single move would involve a quite complex compromise based on predictions about winds and game movements plus, no doubt, other factors now beyond our grasp.

Archaeological Implications of the Model

The goal of this study is to test the validity of two basic assumptions: 1) modern wind patterns resemble those of the last several centuries, and 2) the upper Seacow Bushmen allowed wind direction to partly determine where they camped. If both assumptions are valid, then protected Smithfield camps should be sheltered *more frequently* from the four directions summarised above, and less frequently from

Putting the Wind Up the Smithfield: Seasons of Occupation

southwesterly and northwesterly positions. Windshelter directions in a large enough sample of measured camps should be markedly nonrandom. Furthermore, the rank order of preferred shelter should be: NW-W sheltered camps most numerous in the sample, followed by N-NNW, followed by SE-ESE followed by S.

Practical complications inevitably intervene. Sites in the study sample are not all protected by the same number of vectors (Fig. 12.15), and sites with different amounts of protection cannot be compared in the same statistical test. This means that the sample must be subdivided into nine categories based on the number of contiguous protecting vectors, and only three of these are large enough to furnish statistically valid tests for randomness.

The Test Case

Each class is discussed briefly in turn:

1-Vector sites (n=32) These are mostly small sites backed up against isolated dolerite outcrops or low promontories. Nine other 1-vector sites were excluded from the sample because the main protection was a large bush rather than a rock. One-vector sites are somewhat ambiguous in this test because they imply either a choice of shelter from very specific directions or spurious "shelter" provided coincidentally at a site which was chosen for other reasons (view, surface, etc.). The distribution around the 16-vector array of compass points is clearly nonrandom (Fig. 12.16). NW heads the ranking, as predicted, and there are sites sheltered from all the other predicted directions, but there are some anomalous vectors as well. Windshelter fails to explain the four SW- or the two NE-protected sites. Except for these, the predicted rank order holds for the other sites. The anomalies hint that wind was not always the sole cause of camp choice, at least in this class.

2-Vector sites (n=7) are also valuable but inexplicably scarce. There is one NNW-NW, two WSW-SW, two SW-SSW, a S-SSE, and a NNE-N. The bias towards southwesterlies may be sampling noise, but note that this group was quite well represented in the 1-vector sites also. While it may be argued that the two WSW-SW were actually sheltered from south-veering spring westerlies, and that the two SW-SSW were sheltered from rare winter southerlies, this approaches special pleading and does little to strengthen the case.

3-Vector sites (n=33) are clearly nonrandom in their directional distribution, with NNW-NW dominant and a good fit with the predicted model (Fig. 12.17). Five anomalous southwesterlies once again prevent a perfect fit, and the predicted rank order is also disrupted by three E-NE sites. Sample-size noise may obscure the significance of these anomalies, however.

4-Vector sites (n=15) also display nonrandom distribution, again with most sites in the N-W vectors (Fig. 12.18). Three northeasterlies spoil the fit with the predicted rank order, but sampling noise again interferes.

5-Vector sites (n=63) provide the largest sample and consequently the most reliable test. Five-vector protection is typically obtained by camping at the foot of a dolerite dyke or on level terraces of a hill. In most places, it is the most complete shelter to be found. Nonrandom distribution is obvious (Fig. 12.19), but the vector spread is now so wide that some sites may be protected from two different directions. Thus the dominant group is protected from NW-SW. The next in rank order is the SW-SE group, then the NE-NW group, then SE-NE. This does not fit the predicted rank order because the S- and SE-protected sites are merged in one class. SW-sheltered sites are notably absent in this well-sheltered class, suggesting that their persistent low numbers in the 1–3 vector classes may be spurious noise.

6-Vector sites (n=6) occur in rare shallow alcoves typically at the base of sinuous dolerite outcrops or shallow siltstone cliffs. There are two NNW-SW, a NW-SSW, a W-SSE, a SW-ESE, and SSW-E. Each offers generalised protection around a central vector, respectively: W, W, SW, S, and SE. Note that some of these are wide enough to cope with two of the wind types predicted in the model. It is hardly surprising, therefore, that these loci harbor large, dense sites. The single "anomalous" SW-centered site also received shelter from S, and is not necessarily anomalous.

7-Vector sites (n=8) are in similar sorts of settings and can handle two or more prominent wind directions. There is a NNW-S, a W-SE, a SW-E, three S-NE, a SE-N, and a N-SW. Central vectors are, respectively: WSW, SSW, SSE, ESE, ENE, WNW. None of these need be considered anomalous.

8-Vector sites (n=1) Deep alcoves are very rare. This setting covers NW-SSE, with central vectors backing into SW-WSW, but wide enough to cover complete winter swings in wind direction.

9-Vector sites (n=10) occur typically in the amphitheaters and embayments of dolerite hill swarms. Although ideal for multidirectional shelter, such places resemble fly-blown ovens in summer! Although a small sample, the distribution is markedly nonrandom. There is one NNW-SSE, four NW-SE, two S-N, and three NE-SW. Central vectors are WSW, SW, and NW, respectively. Again, the SW-centered specimen is so wide that it need not be regarded as anomalous.

To summarise, the distribution of windshelter directions is distinctly nonrandom in all classes, including those with very small samples. Further, the dominant vectors in all the classes are NW-W. So far, this result fits the predictive model quite well. It seems reasonable to suggest that assumption (2) is confirmed on this first test, namely, Bushmen allowed wind direction to influence their decisions about where to camp. However, the predicted rank order of frequency: [(1)NW-W, (2)N-NNW, (3)SE-E, (4)S] can only be fitted to the 1–4 vector categories. The 5–9 vector classes are too wide so that two or more vectors in the model become mixed. The 5-vector sites yield the most reliable sample, but mixing spoils the perfection of fit. The persistence of small numbers of SW-protected sites in the 1–3 vector class may be a spurious coincidence because windshelter was not always a key factor in the decision to camp here. At present there are no means for testing this suggestion, however. There remains a faint possibility that past wind patterns included more southwesterlies, so that assumption (1) is not wholly confirmed. This at at least can be further tested by enlarging the sample in future field seasons.

Discussion

Obviously, independent testing by conventional means is essential to crosscheck these results. Isotope analysis cannot be contemplated because human skeletal remains have proven extremely difficult to locate. Evidently the dead were buried at great depth in riverbank silts, without any surface markers. Our entire survey has only located two burials, both truncated by donga erosion.

Two other lines of investigation suggest themselves. The first is a study of the seasonal availability of food plants, about which nothing is presently known. Although we have learned to recognise over a dozen edible tubers, seeds, and berries growing in the area, botanical identifications of collected specimens are still awaited at the time of writing. Nothing is yet known of their nutritional status, yields, or natural history. If their seasonal distribution is the basic "drive mechanism" of the upper Seacow mobility pattern, then a study of present-day patterns of availability *may* yield analogs of past availability, if modern sheep farming practices have not totally disrupted their normal distribution.

The second line of testing will be by conventional methods. There are dozens of small rock shelters in the study area discussed in this paper, and in other parts of the surveyed valley. All have shallow, implementiferous fills which will certainly yield faunal remains, and perhaps some seeds. Other plant remains have not been recovered from the two shelter deposits thus far excavated. Analysis of these assemblages is still in progress, and it remains to be determined whether any unambiguous seasonal indicators will be forthcoming. If any can be found, their presence/absence and frequencies at a given rock shelter can be tested against the direction of windshelter provided by the rear wall. As available shelter orientations cover the whole gamut of angles, a relatively rigorous test would become available.

Summary

Conventional methods for establishing the season of occupation at archaeological sites have been called into doubt by recent isotopic studies of associated human skeletons. In this chapter, the potential of windshelter analysis is explored as an alternate independent test of site seasonality, where skeletal material is unavailable. The underlying assumptions and spatial limitations of the method have been spelled out and procedures are outlined. Very abundant archaeological surface residues (the Smithfield Industry) of the upper Seacow River Bushmen, South Africa, were used as a test case. Wind pattern records from a station near the study area were analysed, and a predictive model of seasonal windshelter was inferred. A fairly good fit was obtained between existing wind patterns and windshelter data from 175 Smithfield camps. Thus, windshelter analysis may be regarded as a very promising approach to seasonality studies, given the appropriate setting and weather data. Proof of its potential must await extensive further testing.

Acknowledgements

ZVAP research has been supported by the National Science Foundation; by the Wenner-Gren Foundation; by Southern Methodist University; by the Institute for the Study of Earth and Man at SMU; by the University of Cape Town; and by the National Museum, Bloemfontein. My wife Beatrix has maintained the entire logistical support of the project. The field data were collected by Tim Hart, Conrad Steenkamp, and Reg Webster. Statistical tests were carried out by Britt Bousman. I also wish to thank collectively the personnel of the ZVAP survey team, and the many Karoo farmers who have assisted and befriended us over the years. They are named individually in Sampson (1985). Special thanks are due to Dr. Piet Roux, Director of the Grootfontein College of Agriculture and Research Institute, Middelburg, South Africa, for his interest and support, and to Dr. Louis Botha, resident meteorologist at Grootfontein for providing raw data and advice.

References

Barnard, A. 1979. Kalahari Bushman settlement patterns. In *Social and Ecological Systems* (eds P. Burnham and R. F. Ellen): pp. 131–44. London: Academic Press.

Batschelet, E. 1981. *Circular Satistics in Biology*. London: Academic Press.

Beaumont, P. B. and Vogel, J. C. 1984. Spatial patterning of the ceramic Later Stone Age in the northern Cape Province, South Africa. In *Frontiers: Southern African Archaeology Today* (eds M. Hall, G. Avery, D. M. Avery, M. L. Wilson and A. J. B. Humphreys): pp. 80–95. Cambridge: BAR International Series 207.

Brooks, A. S. 1984. San land-use patterns, past and present: Implications for southern African prehistory. In *Frontiers: Southern African Archaeology Today* (eds M. Hall, G. Avery, D. M. Avery, M. L. Wilson and A. J. B. Humphreys): pp. 40–52. Cambridge: BAR International Series 207.

de Lumley, H. 1969. *Une cabane Acheuléenne dans la Grotte du Lazaret (Nice)*. Tome 7. Paris: Société Préhistorique Française.

Fletemeyer, J. R. S. 1977. Age determination in the teeth of the Cape Fur Seal and its bearing on the seasonal mobility hypothesis proposed for the western Cape, South Africa. *South African Archaeological Bulletin* 32:146–9.

Hitchcock, R. K. 1982. Patterns of sedentism among the Basarwa of eastern Botswana. In *Politics and History in Band Societies* (ed E. Leacock and R. Lee): pp. 223–67.

Mobley, C. M. 1981. *Archaic hunter-gatherer settlement in Northeastern New Mexico*. (Ph.D. dissertation, Southern Methodist University). Ann Arbor: University Microfilms International.

Monks, G. G. 1981. Seasonality studies. In *Advances in Archaeological Method and Theory Vol. 4* (ed M. B. Schiffer): pp. 177-240. New York: Academic Press.

Ott, L. 1977. *An Introduction to Statistical Methods and Data Analysis*. N. Scituate, Mass.: Duxbury Press.

Parkington, J. E. 1972. Seasonal mobility in the Late Stone Age. *African Studies* 31:223–43.

_____. 1984. Changing views of the Later Stone Age of South Africa. In *Advances in World Archaeology vol. 3* (eds F. Wendorf and A. Close): pp. 90–142. London: Academic Press.

Partridge, T. C. and Dalbey, T. S. In press. Geoarchaeology of the Haaskraal Pan: A preliminary palaeoenvironmental model. *Palaeoecology of Africa 17*. Rotterdam: Balkema

S. A. Department of Transport, Climatology Branch of the Weather Bureau. 1960. *Climate of South Africa part 6: Surface winds*. Pretoria: The Weather Bureau, Union of South Africa.

Sampson, C. G. 1972. The Stone Age industries of the Orange River Scheme and South Africa. *Memoirs of the National Museum, Bloemfontein* 5:1-288.

_____. 1974. *The Stone Age Archaeology of Southern Africa*. London: Academic Press.

_____. 1984. Site clusters in the Smithfield settlement pattern. *South African Archaeological Bulletin* 39:5–23.

_____. 1985. Atlas of Stone Age settlement in the central and upper Seacow valley. *Memoirs of the National Museum, Bloemfontein* 18:1–110

_____. In press. *Stylistic Boundaries among Mobile Hunter-Foragers*. Smithsonian Institution Press.

Sealy, J. C. and van der Merwe, N. 1985. Isotope assessment of Holocene human diets in the southwestern Caope, South Africa. *Nature* 315:138–40.

Tyson, P. D. 1969. Atmosphere circulation and precipitation over South Africa. *Environmental Studies, Occasional Paper* 2:1–22.

van der Merwe, P. J. 1937. *Die Noordwaartse Beweging van die Boere Voor die Trek (1770–1842)*. Pretoria.

Putting the Wind Up the Smithfield: Seasons of Occupation

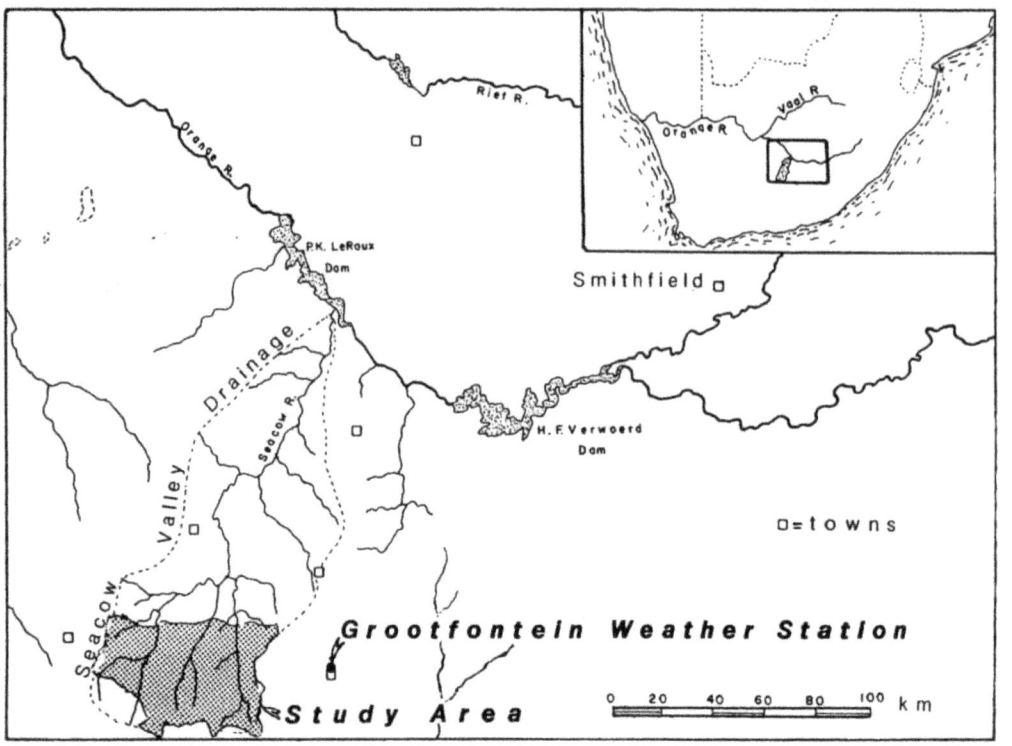

Fig. 12.1: The weather station is located only 29 km from the eastern rim of the study area, and 87 km from its western rim. Station data are appropriate for modelling weather patterns in the study area, therefore.

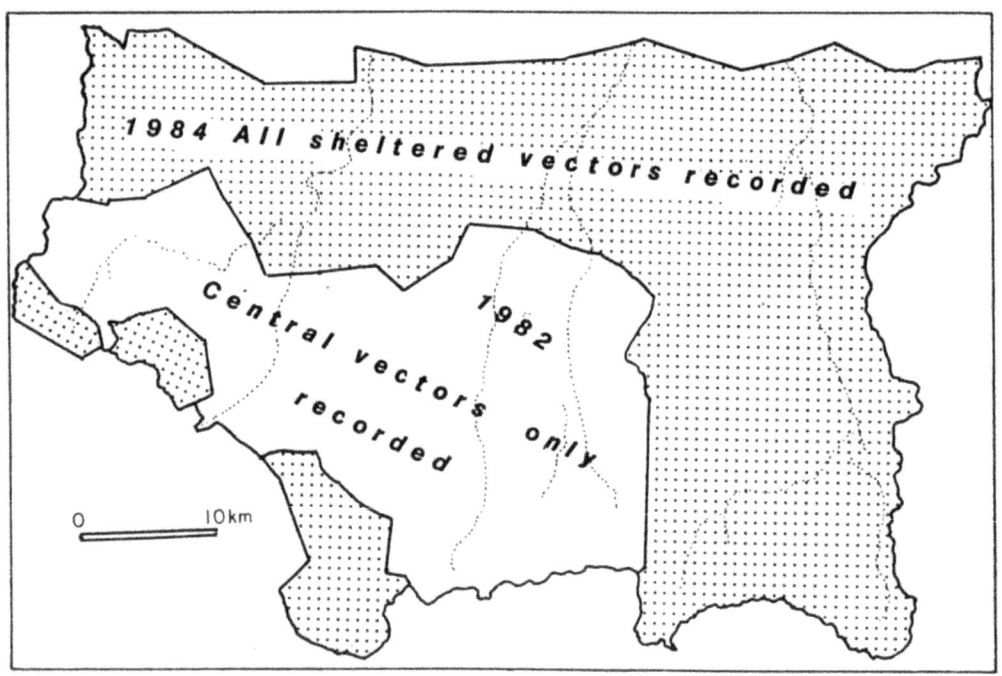

Fig. 12.2: Windshelter data were gathered during two seasons. Only the 1984 data are analysed in this study.

C. G. Sampson

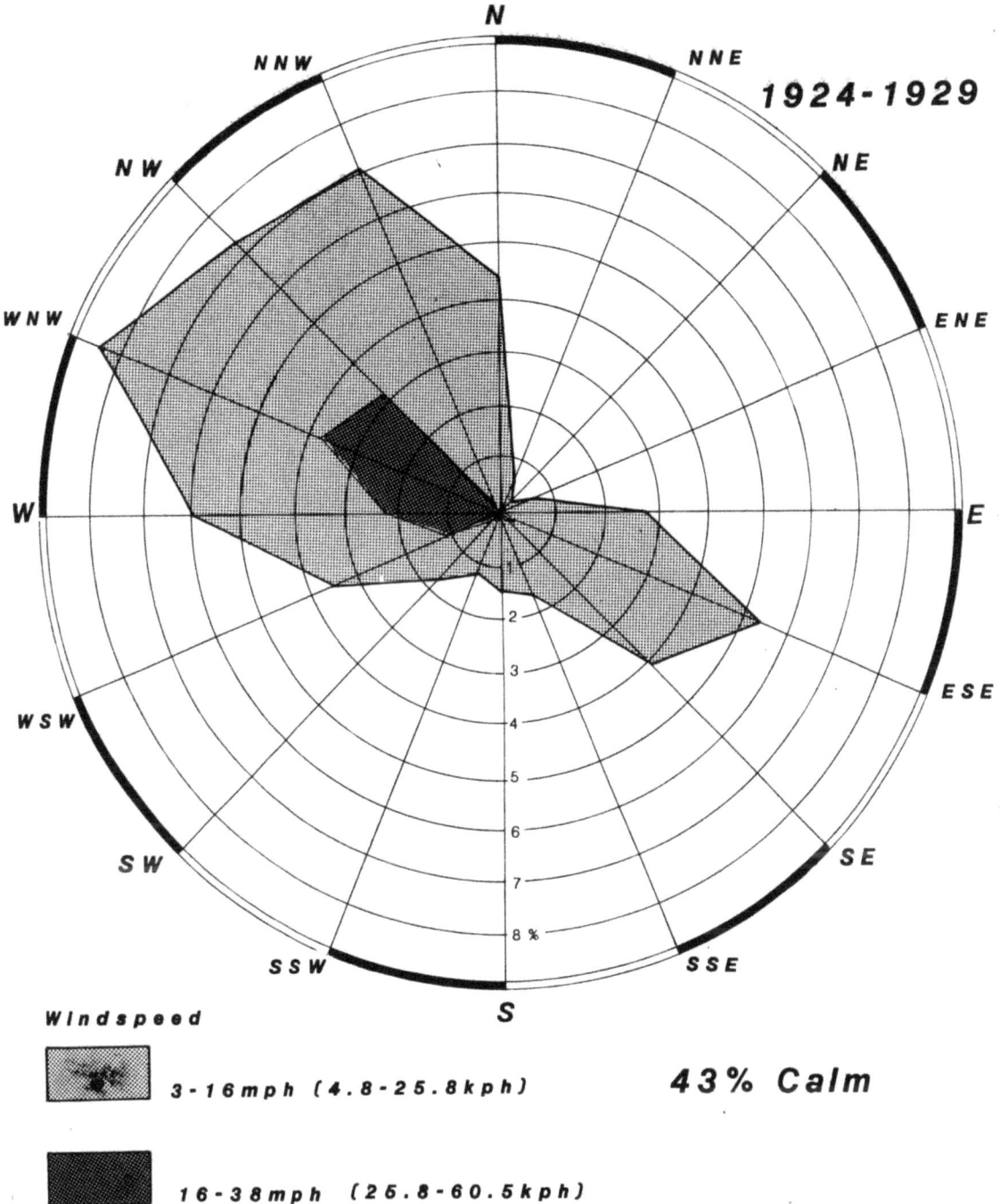

Fig. 12.3: The Mean Annual Windrose reflects the duration of two classes of wind velocity, averaged over 5 years. Winds are bi-directional, with a clear bias toward the N-WSW vectors. Most of the winds blow during daylight hours (e.g., Fig 12-14) with more calm air at night. Constructed from published data (S.A. Dept. of Transport1960). The wind gauge then in use was a sprung brass baffle plate (P. Botha, personal communication) marked with relatively few velocity intervals, calibrated in mph.

Putting the Wind Up the Smithfield: Seasons of Occupation

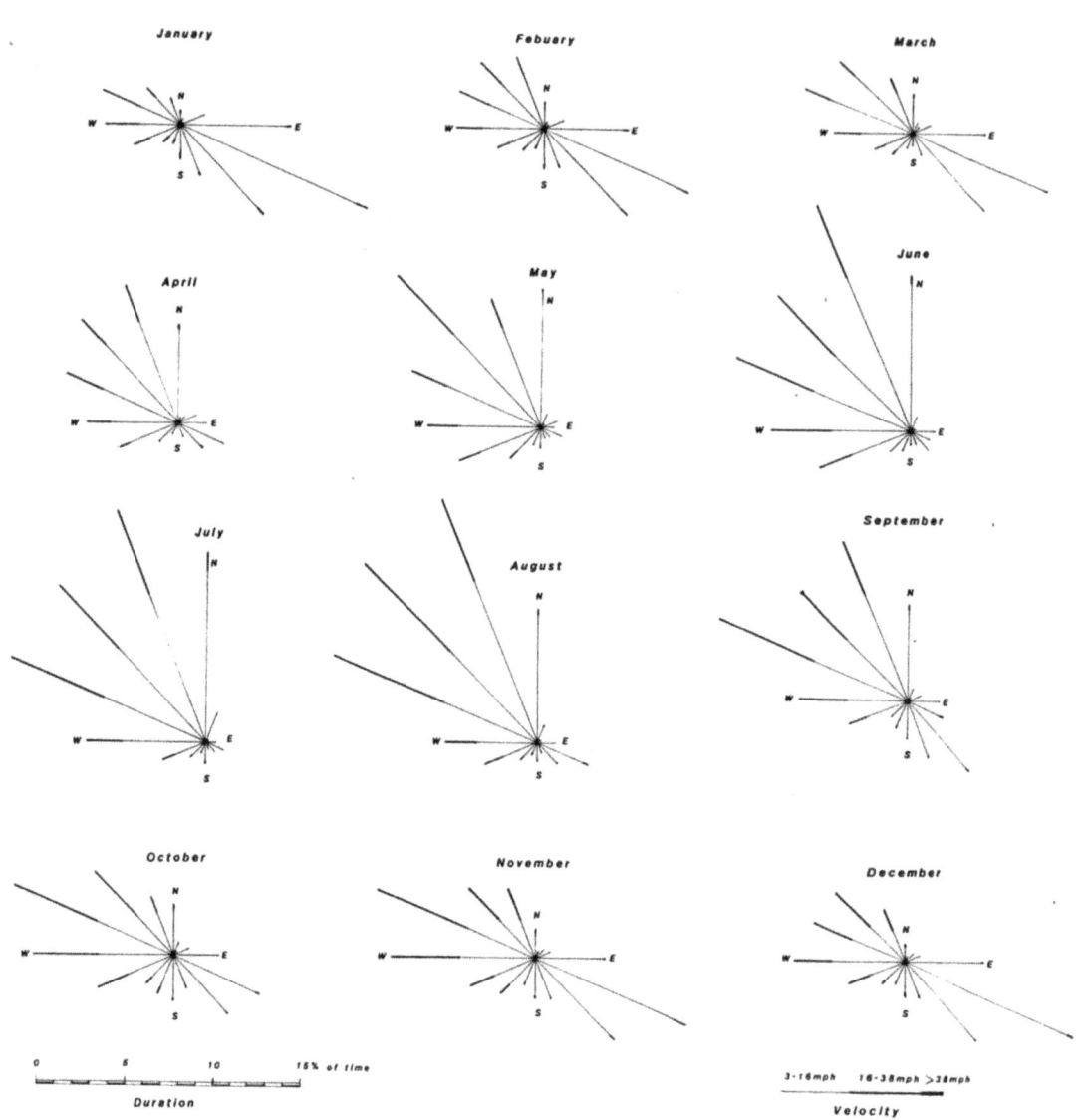

Fig. 12.4: Mean Monthly Windroses show N-WSW dominant in late winter/early spring, and SE-E dominant in summer.

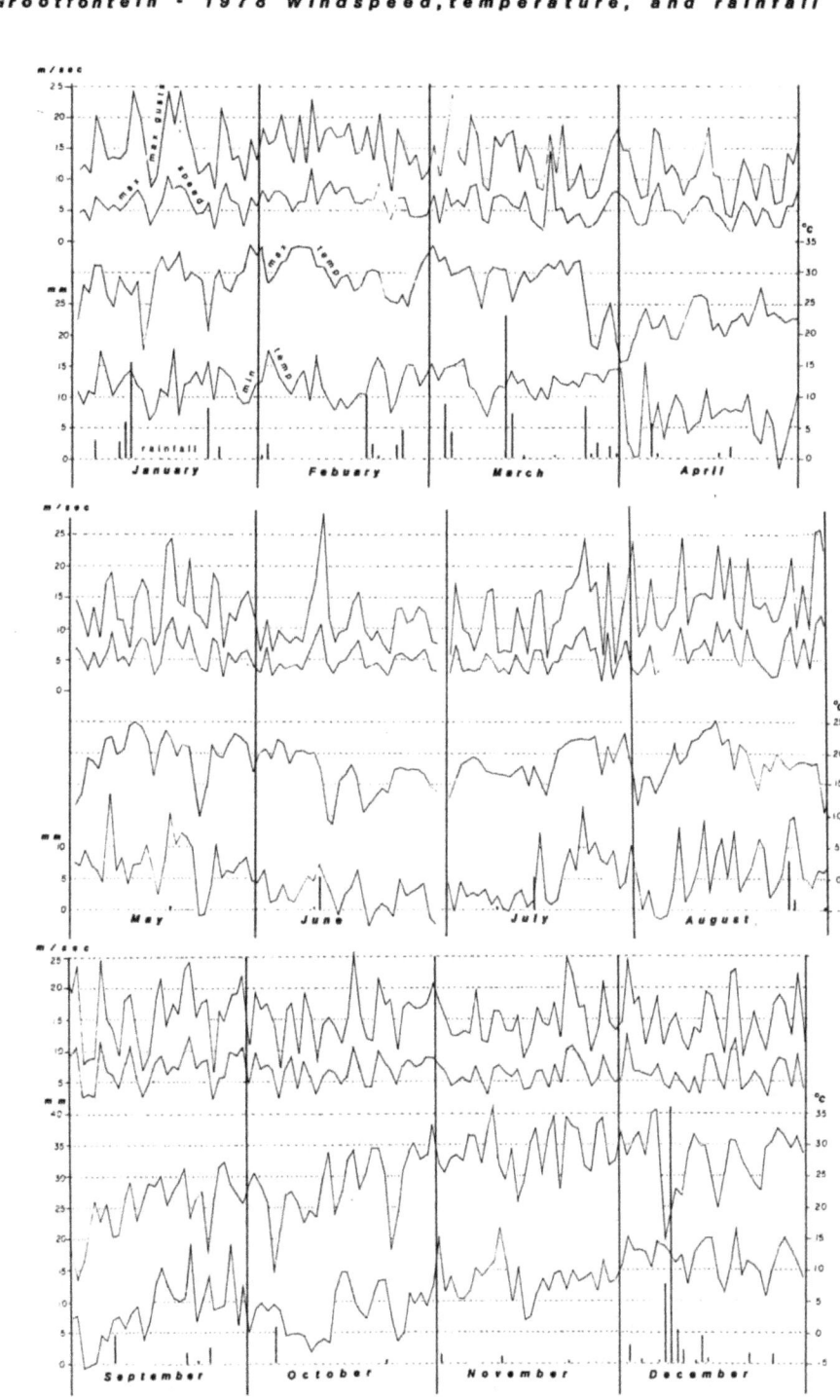

Fig. 12.5: Daily extremes of wind velocity and temperature show windy days (mostly NNW-W) are associated with warmer temperatures. Calmer, cooler days are associated with southeasterlies.

Putting the Wind Up the Smithfield: Seasons of Occupation

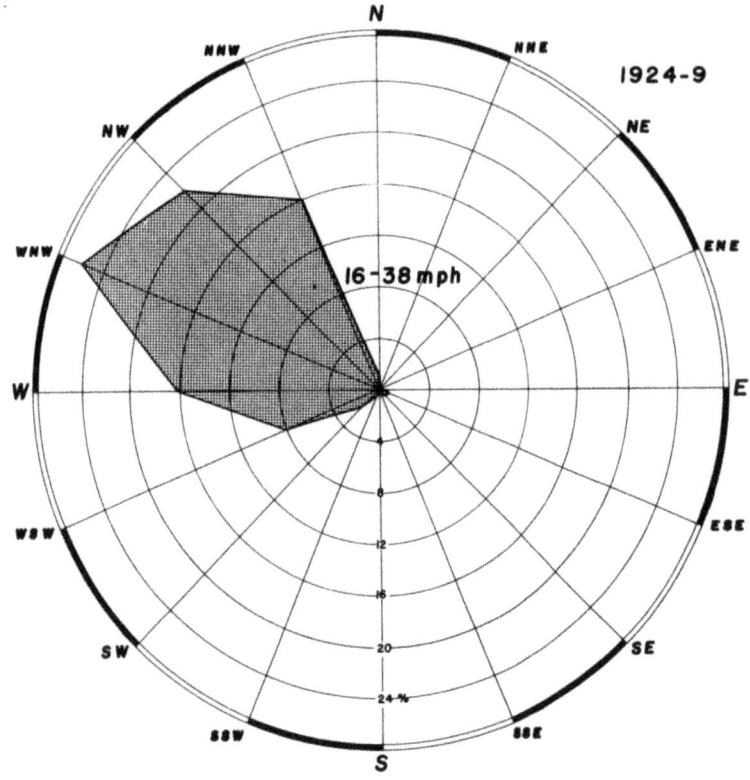

Fig. 12.6: High velocity winds blow mainly from the NNW-WSW vectors.

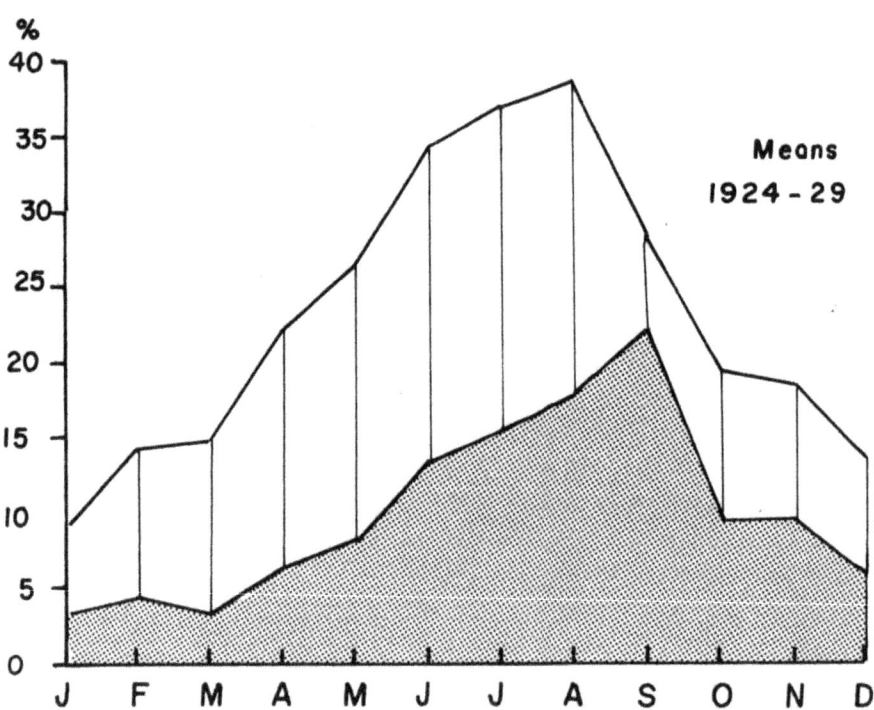

Fig. 12.7: Mean monthly duration of winds from NNW-WNW vectors combined. July-August are the windiest months, but the 16-38mph (stippled) shows that September has the worst gusts, including 0.4% at velocities greater than 38 mph, not shown on this graph.

C. G. Sampson

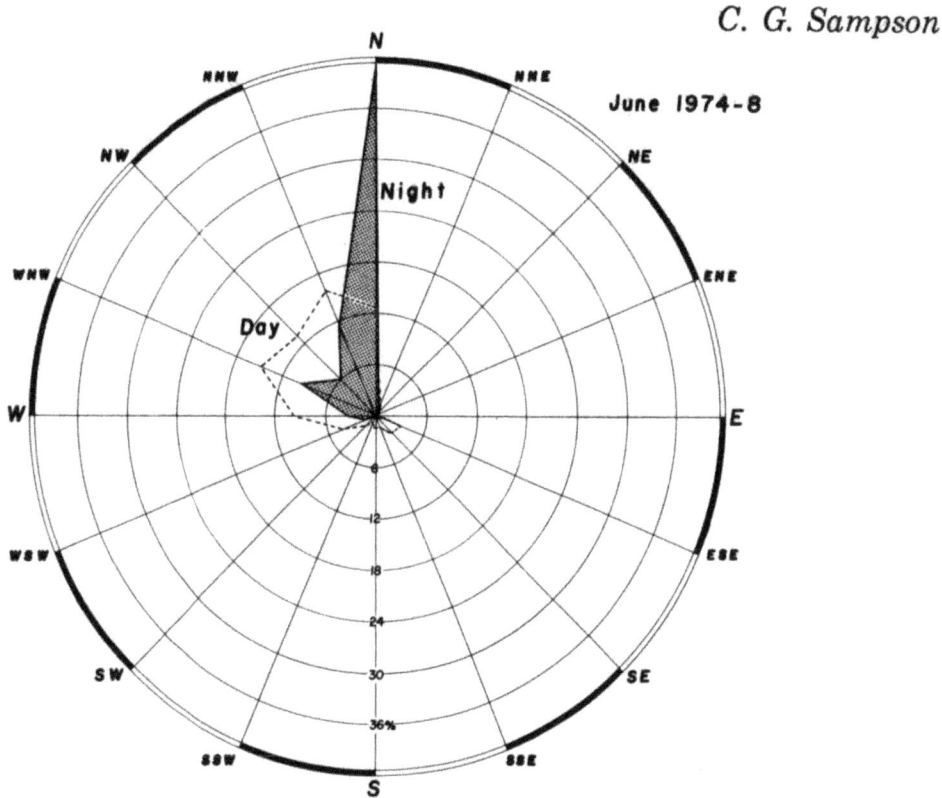

Fig. 12.8: Contrasts between day (07-18 hrs) and night (19-06 hrs) wind directions in midwinter. Nights average 22% calm. Days average 16% calm. Freezing night winds blow mainly from N-NNW vectors.

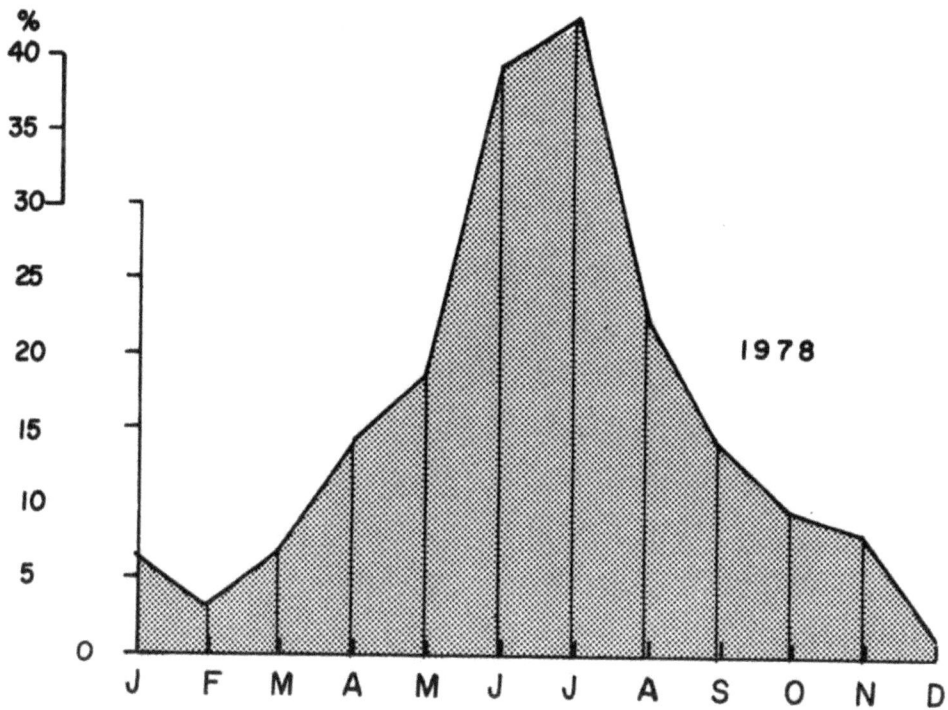

Fig. 12.9: Mean monthly duration of all North wind blowing at night (19-06 hrs) during 1978. Nighttime northers are most common in winter.

Putting the Wind Up the Smithfield: Seasons of Occupation

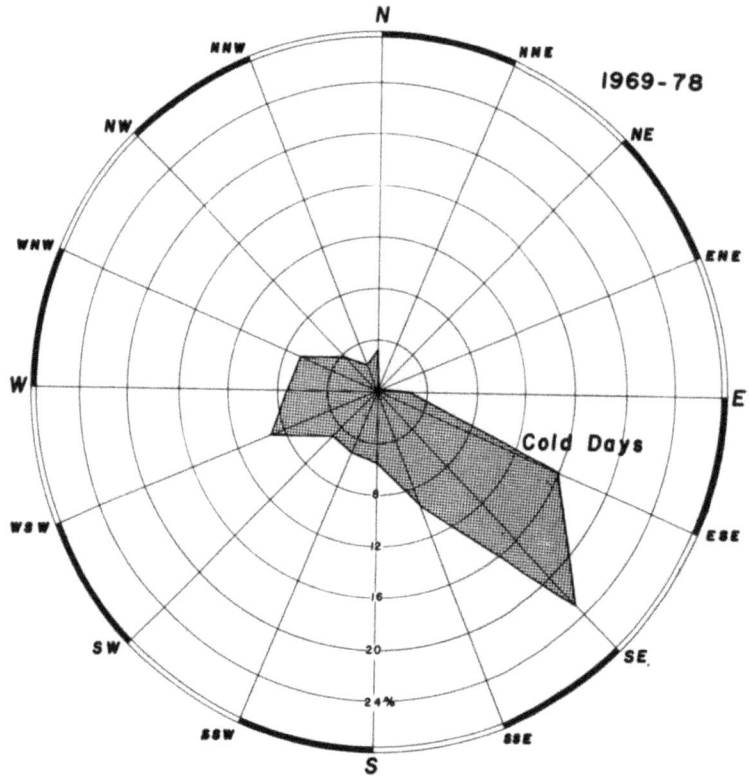

Fig. 12.10: Mean duration of all winds blowing in days (07–18 hrs) when the maximum temperature failed to rise above 10° C. Cold days are strongly associated with winds from the SSE-ESE vectors.

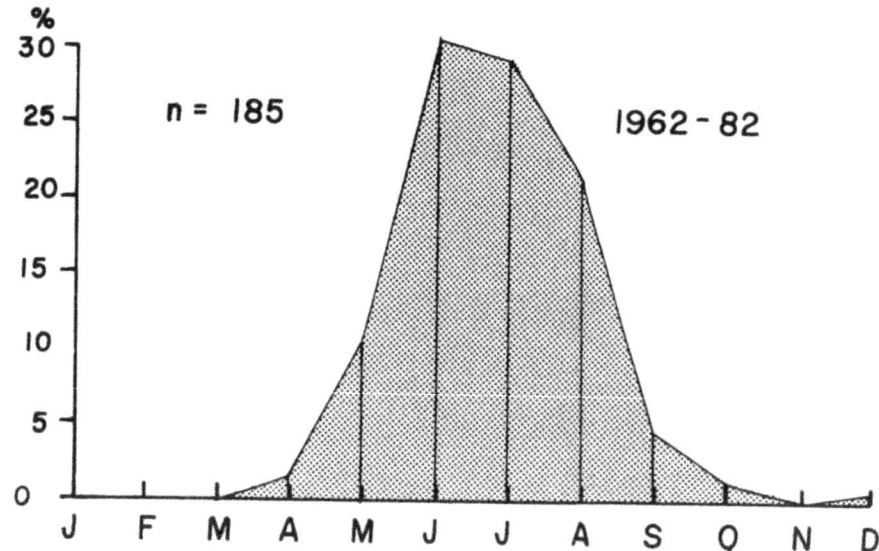

Fig. 12.11: Monthly distribution of cold days (maximum temperature below 10° C). There are, on average, only 9.25 cold days per year.

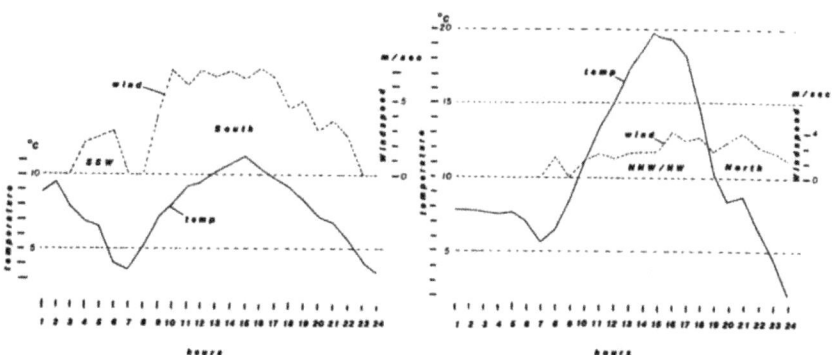

Fig. 12.12: Hourly windspeeds and temperatures for May 1, 1978 (*left*) and two days later (*right*). Note how the south wind holds down daytime temperatures and how the nighttime norther depresses temperature.

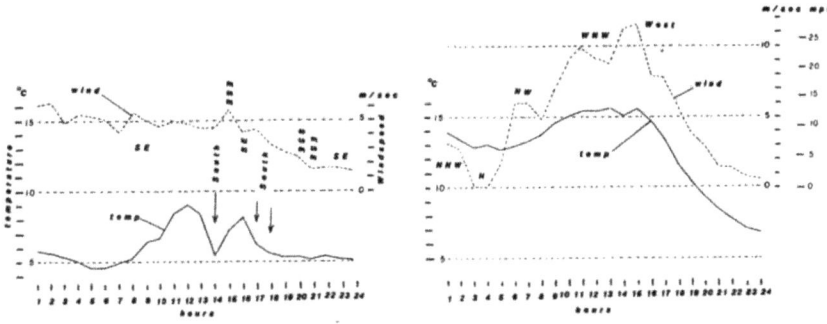

Fig. 12.13: Hourly windspeeds and temperatures for October 5 (*left*) and September 21, 1978 (*right*). When southeasterlies veer to south on cold days (*left*) temperature drops rapidly and sharply. The high velocity northwesterlies (*right*) bring warmer temperatures.

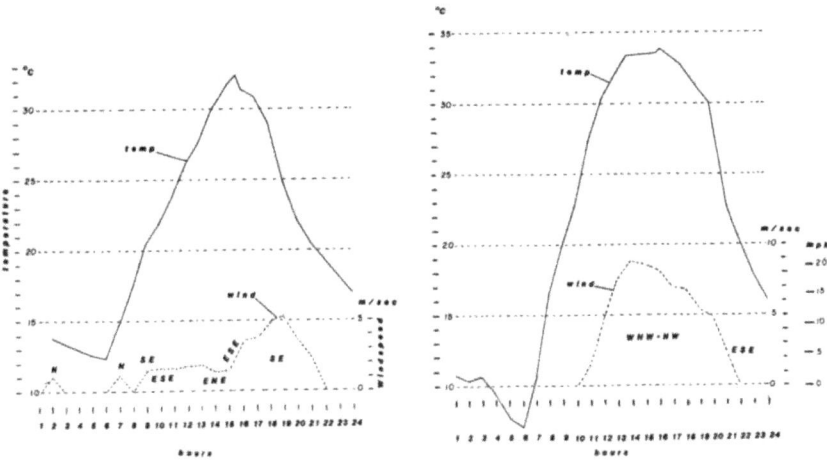

Fig. 12.14: Hourly windspeeds and temperatures for two typical summer days: January 18 (*right*) and 31, 1978 (*left*). Note the rapid climb in morning temperature in calm air (*right*) leading to a long hot afternoon.

Putting the Wind Up the Smithfield: Seasons of Occupation

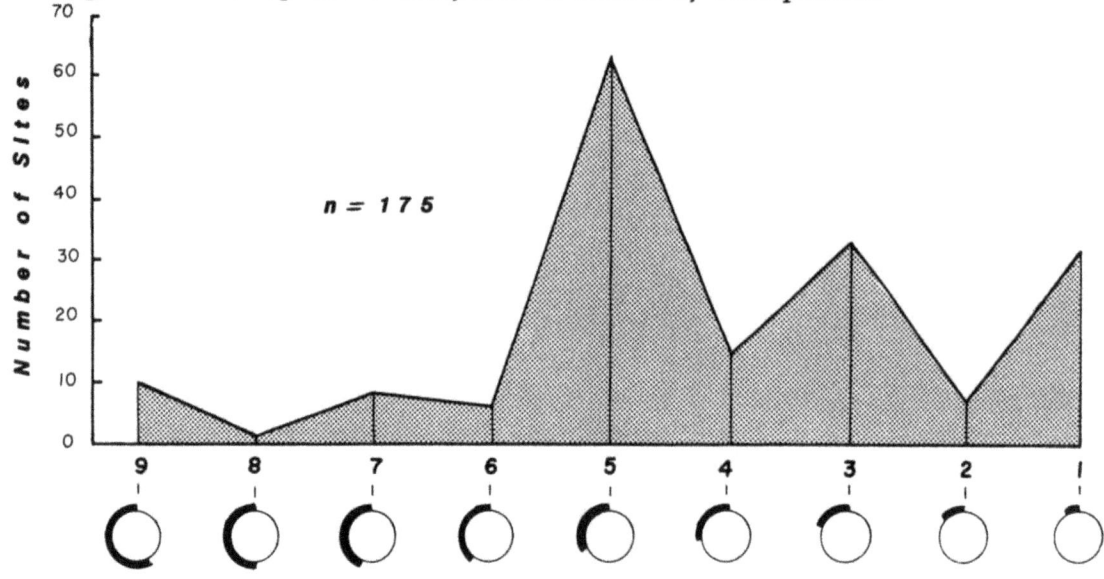

Fig. 12.15: Most sites are protected from 5 vectors. Alcoves and amphitheaters offering 6 or more vectors of windshelter are scarce in this landscape.

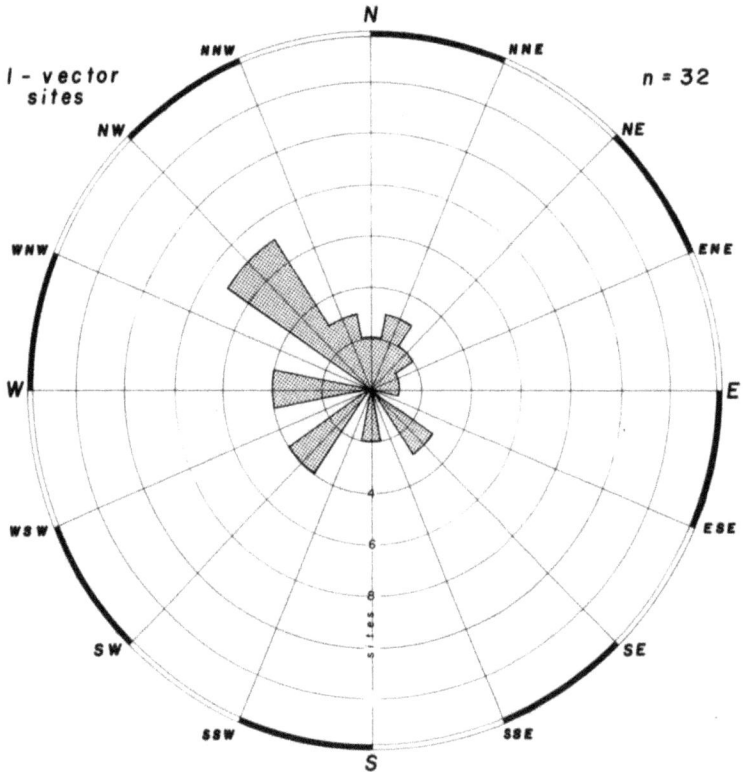

Fig. 12.16: Vector distributions of 1-vector sites. Distributions appear to be non-random, with preferences shown for NW, W and SW. This would also appear to be a good (but not perfect) fit with the windshelter model. The four SW-sheltered sites spoil the supposed fit. The sample size is too small to accept a Chi-square test, however. Also, the classification of data into 16 arbitrary units (rather than angular readings in degrees) prevents the use of one of several circular statistical tests applicable to samples of this size (Batschelet 1981).

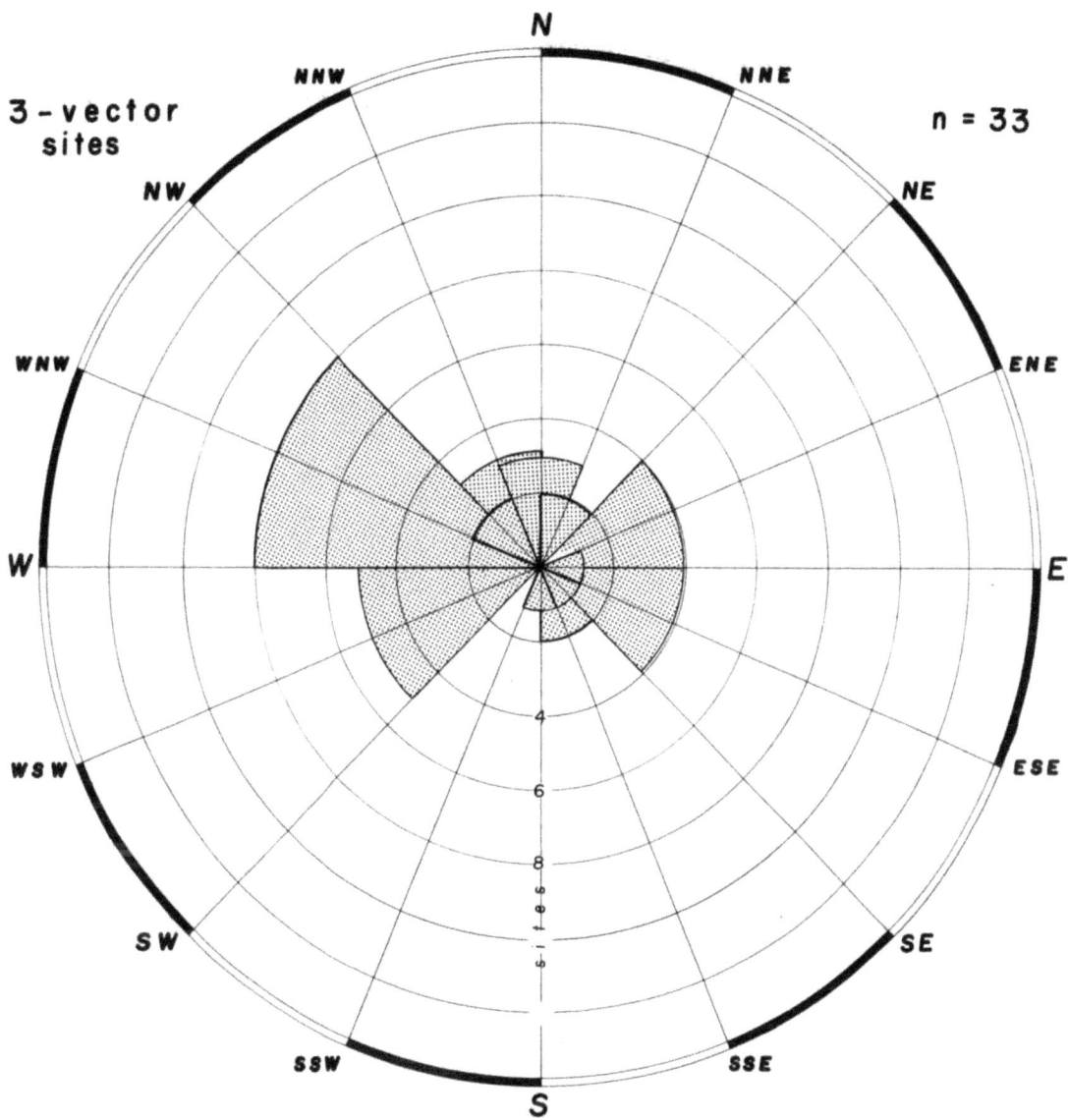

Fig. 12.17: Vector distribution of 3-vector sites. The apparently non-random distribution suggests a bias in favor of NW-W. However, five W-SSW sites may indicate an imperfect fit with expectations of the model. It remains impossible to determine whether or not these biases are in fact merely the result of random chance, as the sample size is too low for a valid test of the null hypothesis.

Putting the Wind Up the Smithfield: Seasons of Occupation

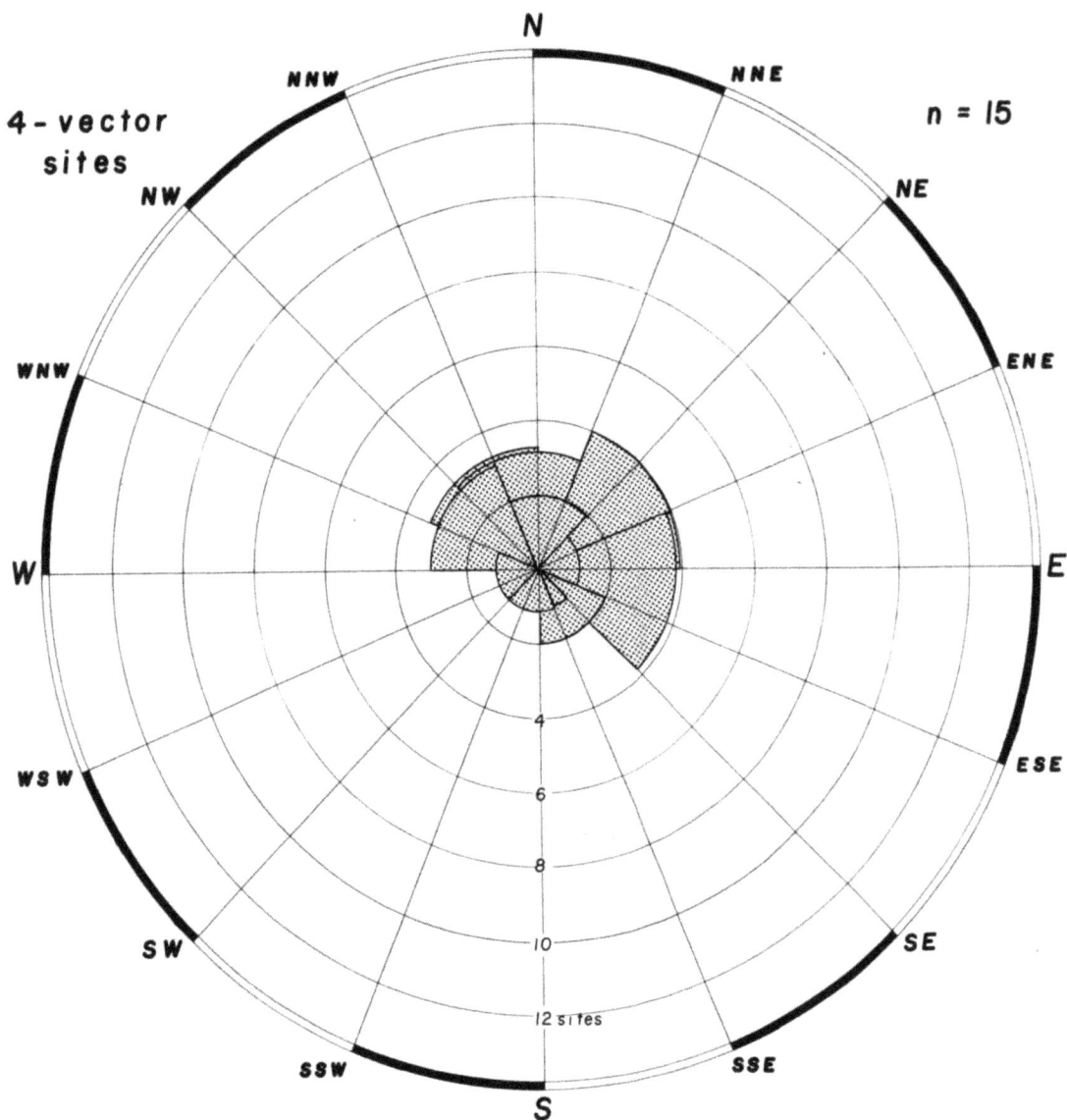

Fig. 12.18: Vector distribution of 4-vector sites. A near-perfect fit with the expectations of the windshelter model is suggested, but the four northeasterlies may spoil this comparison. Again, the sample is too small for a valid test of the null hypothesis.

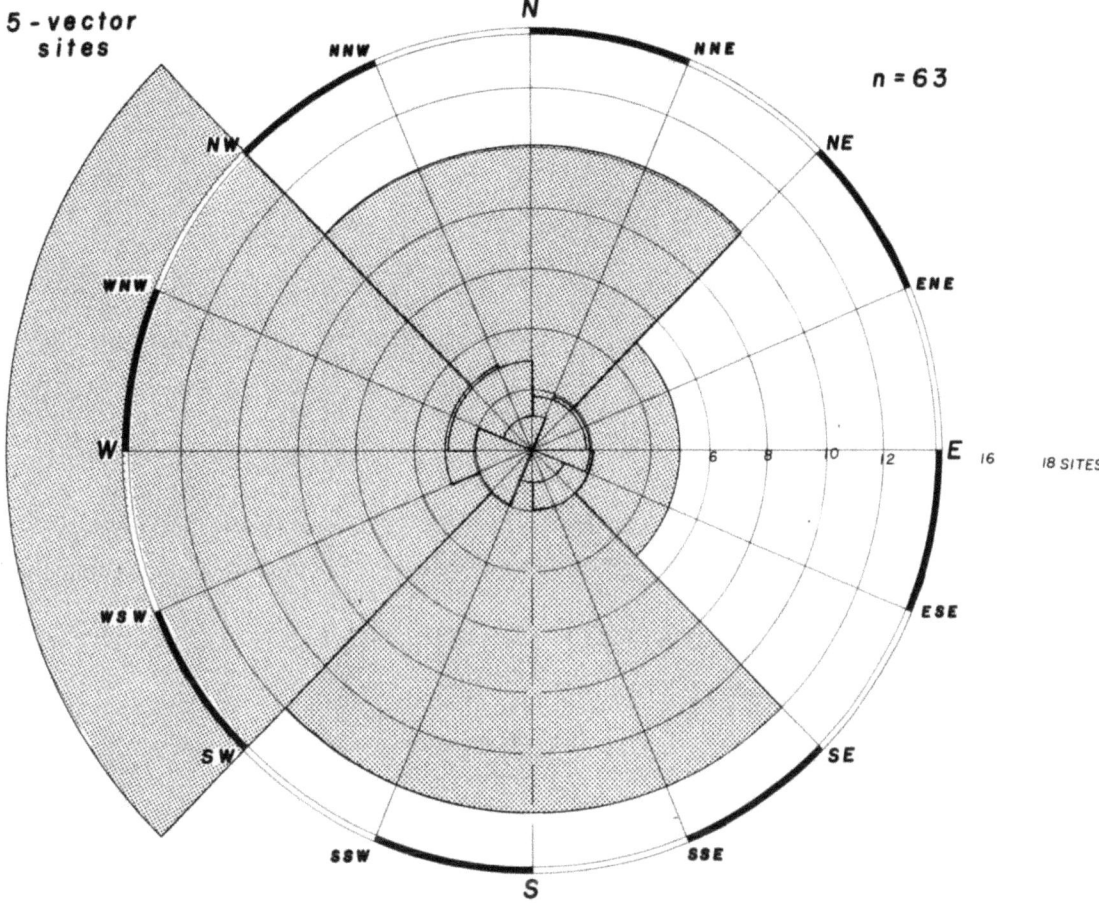

Fig. 12.19: Vector distribution of 5-vector sites. There appears to be a good general fit with the expectations of the model, but the extent of windshelter is now too wide, so that a single site can encompass more than one direction in the model. Although the sample-size is still rather too low to test the null hypothesis, it is nevertheless the largest sample available. A Chi-square Goodness-of Fit Test was calculated, notwithstanding. The null hypothesis expects an even distribution of sites in all 16 classes, i.e., 3.9375 sites per class. The test result (X^2=97.796, df=15, p value=<0.0001) is highly significant, however. Even though the expected values are too low to meet the assumptions of the test (e.g., Ott 1977), the observed distribution deviates enough from the expected to be truly significant. The X^2 test cells contributing most to the deviation are centered on W, S, and N. When the centers of *all sites* in the entire sample are distributed around 32 vectors, a (valid) test of the null hypothesis yields: X^2=164.06; df=31; p value <0.0001. The difference is again highly significant, with most of thé deviation contributed by NW, W, S, and N.

www.ingramcontent.com/pod-product-compliance
Ingram Content Group UK Ltd.
Pitfield, Milton Keynes, MK11 3LW, UK
UKHW060200240426
12048UKWH00029B/1674